RESIDENTIAL CONSTRUCTION ACADEMY

Instructor's Resource Guide to Accompany

Basic Principles for Construction

SECOND EDITION

MARK HUTH
KERRY A. REINHACKEL

THOMSON
DELMAR LEARNING™

Australia• Canada• Mexico• Singapore• Spain• United Kingdom• United States

THOMSON

DELMAR LEARNING

Instructor's Resource Guide to Accompany
Residential Construction Academy: Basic Principles for Construction, Second Edition
by Mark Huth and Kerry A. Reinhackel

Vice President, Technology and Trades ABU:
David Garza

Director of Learning Solutions:
Sandy Clark

Managing Editor:
Larry Main

Senior Acquisitions Editor:
James DeVoe

Senior Product Manager:
John Fisher

Marketing Director:
Deborah S. Yarnell

Marketing Manager:
Kevin Rivenberg

Marketing Specialist:
Mark Pierro

Director of Production:
Patty Stephan

Production Manager:
Stacy Masucci

Content Project Manager:
Jennifer Hanley

Art Director:
Benjamin Gleeksman

Editorial Assistant:
Tom Best

Printed in the United States of America
1 2 3 4 5 XX 09 08 07

For more information contact Thomson Delmar Learning
Executive Woods
5 Maxwell Drive, PO Box 8007,
Clifton Park, NY 12065-8007
Or find us on the World Wide Web at
www.delmarlearning.com

Library of Congress Cataloging-in-Publication Data:

ISBN-10: 1-4180-5252-3
ISBN-13: 978-1-4180-5252-2

NOTICE TO THE READER

Publisher does not warrant or guarantee any of the products described herein or perform any independent analysis in connection with any of the product information contained herein. Publisher does not assume, and expressly disclaims, any obligation to obtain and include information other than that provided to it by the manufacturer.

The reader is expressly warned to consider and adopt all safety precautions that might be indicated by the activities herein and to avoid all potential hazards. By following the instructions contained herein, the reader willingly assumes all risks in connection with such instructions.

The publisher makes no representation or warranties of any kind, including but not limited to, the warranties of fitness for particular purpose or merchantability, nor are any such representations implied with respect to the material set forth herein, and the publisher takes no responsibility with respect to such material. The publisher shall not be liable for any special, consequential, or exemplary damages resulting, in whole or part, from the readers' use of, or reliance upon, this material.

Table of Contents

Chapter Outlines

Appendices

Note to Instructors

The *Instructor's Resource Guide* to accompany *Residential Construction Academy: Basic Principles for Construction,* Second Edition is designed to facilitate classroom preparation for you as the instructor. It is divided into two sections: 1) *Chapter Outlines* and 2) *Appendices.*

- The *Chapter Outlines* provide a guide to the major concepts presented in each chapter in the textbook. A basic outline helps to ensure that you hit these major topics during classroom discussion. Correlating PowerPoint presentation slides are outlined in the margins, as well as "Teaching Tips" that provide helpful hints from an experienced instructor. The PowerPoint presentation slides are available on an accompanying e.resource CD-ROM (see below).
- The *Appendices* provide valuable information to instructors as they prepare for classroom instruction:
  — *Appendix A* provides answers to the review questions that appear at the end of each chapter in the book.
  —*Appendix B* provides a quick reference to corresponding PowerPoint slides that appear in the lecture outlines. This allows instructors to highlight particular topics of discussion if they choose.
  —*Appendix C* provides answers to the workbook.

Also available: An accompanying e.resource CD-ROM containing the following:

- *PowerPoint presentations* outlining each chapter in the book provide you with a quick resource for classroom lecture
- *Testbank* in *ExamView 4.0* containing approximately 400 questions to help you prepare exams for your class—completely editable so that you may add or remove questions
- *Image Library* including all photos and line drawings from the text to enable you to create your own presentations for the classroom
- *Instructor's Resource Guide* available in PDF format

***Order #:* 1-4180-5253-1**

Click on www.residentialacademy.com for further information on other learning materials within the *Residential Construction Academy Series.*

Chapter 1 Organization of the Industry

Instructor's Note

In this column, instructors will find a variety of lecture tips to aid student comprehension and notes on when to use PowerPoint slides to teach concepts.

Appendix B lists corresponding numbered PowerPoint Presentation titles

SLIDE 1-1
OBJECTIVES

Teaching Tip: Any construction experience you may have will be invaluable as you progress through this chapter. You can make vocabulary words come to life with your personal experience stories of on-the-job situations.

OBJECTIVES

- List and describe several potential careers in construction.
- Explain the roles of architects, engineers, city building officials, and contractors.
- Describe the major forms of business ownership and the differences between them.
- Explain what a building code is.

Glossary of Terms

Apprentice—a person who is being trained to work in the building trades. Apprentices attend classes and work under the supervision of a skilled craftsman.

Contractor—the person who owns the construction business. Contractors enter into contracts with customers to do specified construction work. Contractors hire workers or other subcontractors to complete the contracted work.

Corporation—a form of business ownership in which people that are not involved in operating the business own shares of the company. A board of directors operates the company.

Craft—see skilled trades.

Developer—the person or company that buys undeveloped land and works with designers and contractors to develop it into more valuable property.

Journeyman—a skilled craft worker who has completed an apprenticeship or otherwise proven his or her ability in the trade.

Laborer—an unskilled or semiskilled worker on a construction site.

Model code—a suggested building code intended to be adapted as is or with revisions to become an official code of a particular government.

Partnership—a form of business in which more than one person shares the ownership and operating duties for a company.

Profession—an occupation that requires more than four years of college and a license to practice.

Semiskilled labor—workers with very limited training or skills in the construction trades.

Skilled trades—carpenters, electricians, plumbers, painters, etc. These occupations require training and skill. The skilled trades are often referred to as the crafts.

Sole proprietorship—a business whose owner and operator are the same one person.

Subcontractor—a contractor who is performing work for another contractor.

Unskilled labor—workers with no specific training in the construction trades. This term also applies to work that does not require training.

A. Residential Construction

SLIDE 1-2

1. One of the biggest sectors of the American economy.
2. One out of every six people is involved in construction in some way. (*Engineering News Record*).
3. Home building accounts for 53 percent of the construction industry.
4. The construction industry offers a variety of occupational opportunities ranging from those who handle the tools to those who draw the plans.

B. Construction Personnel

1. There are four categories of construction occupations:
   a. Unskilled labor or semiskilled labor
   b. Skilled trades or crafts
   c. Technicians
   d. Design and management
2. Unskilled or semiskilled labor
   a. Construction workers with limited skills are called laborers.
   b. Laborers run errands, move materials and work under a skilled worker.
   c. Laborers need to be in great physical condition to do the work.

SLIDE 1-3

   d. Construction laborers are those workers who have not yet reached a high level of skill and are not in a registered apprenticeship program.

Teaching Tip: Emphasize that anyone who wants to can advance in the construction industry, if they are willing to study and work their way up the "ladder" of success.

   e. Laborers may specialize in a particular trade
      - Mason's tender
      - Carpenter's helper
   f. Laborers often go on to acquire skills and become skilled workers.
3. Skilled trades or crafts
   a. An occupation working with tools and materials, and building structures. These are the crafts most directly involved with building construction.

SLIDE 1-4

   b. Apprenticeship programs offered by trade unions, trade associations, technical colleges, and large employers offer opportunities for individuals to learn the skills needed to be employed in the building trades.
   c. Apprentices attend school a few hours a week and work full-time on a job site, under the supervision of a journeyman.
   d. Although the term *journeyman* seems to refer to men, many highly skilled building trades workers are women.
   e. Apprentices earn about half the wage of a journeyman.
      - Wages increase as they complete levels of the apprenticeship program.
   f. Typical apprenticeship
      - Lasts 4 to 5 years
      - 144 hours per year of classroom training
      - 2,000 hours per year of supervised work experience

   g. Weather can make jobs in the building trades challenging, but good workers will always find good jobs in the industry.

4. Technicians

   a. The link between the skilled trades and the professions

SLIDE 1-5
SLIDE 1-6

   b. Technicians need many skills.
      - Mathematics
      - Computer skills
      - Specialized equipment
      - Construction knowledge

This is a good chapter to include guest speakers. If you can get a representative from one or more levels and/or areas discussed in this chapter, it would really be an interesting addition to your lecture. There is nothing quite like seeing success in action.

   c. Work takes technicians from the field to the office.
   d. Community colleges often have construction management courses that provide graduating technicians with a good overview of the construction business.
   e. Wages
      - Start out about the same as skilled workers
      - Can expect more regular employment
      - Opportunities for advancement are better.

5. Design and Management

   a. Professions include architecture, engineering, and contracting.
      - These professions require four years of college.

SLIDE 1-7

      - Individuals usually work in offices.
   b. Contractors are professionals due to the large amount of money with which they deal.
      - May or may not have four years of college
      - Many states require contractor's to be licensed.
   c. Architects' typical educational degrees
      - Bachelor's degree in fine arts
      - Master's degree in architecture
   d. Engineering specialties vary, but civil engineering with a strong background in math and science are more commonly found in construction.
   e. Contractors are the owners of businesses that do most of the building.
   f. The general contractor is the principal construction company hired by the owner to construct the building.
   g. Subcontractors
      - Journeymen who specialize in a field in wihch they have spent a great deal of time
      - Jobs include framing and plumbing, among others
   h. Foremen
      - Come from skilled labor pool
      - Workers with the best ability to lead others
      - Work on specific part of job with small crew

6. Overall View of Design and Construction

   a. Real estate developer

SLIDE 1-8

      - Acquire large tract of land on which he would like to build 1,000 homes

- Land sold later in the development process

b. City planner
- Works with the developer
- Ensures city codes and ordinances are met

c. Civil engineer
- Joins project to meet needs for street planning and house design
- Plans and designs infrastucture
    1) Storm water drainage
    2) Soil conditions
    3) Traffic loads for determination of roadway foundations

d. Architectural firm
- Designs the houses typically using several stock plans
- Allows room for individual small adjustments

e. Architectural plans that involve special structural considerations require a structural engineer to analyze structural requirements and ensure structures will adequately support planned features.

f. Roof trusses require an engineer's expertise.
- Trusses are assemblies that make up the frame of the roof.

SLIDE 1-9

- Designed specifically for the weight of framing materials, weight of the roof covering, anticipated weight of snow, and the span of the truss.

g. End product
- Architect and engineer work out details
- One set of construction drawings
- Drawings require architect's seal

7. Forms of Ownership

a. A construction company's form of ownership may have little to do with the form of ownership.

SLIDE 1-10

b. Sole proprietorship
- Simplest type of ownership
- Owner, proprietor are same person
    1) Advantages
        Complete control
        Minimal government regulation
        High profits for successful companies
    2) Disadvantages
        Owner responsible for any debt
        Owner can be sued for the company
        Owner suffers all losses of the company

c. Partnership
- Similar to a sole proprietorship
- Two or more owners
- Spreads advantages, disadvantages among owners in proportion to each partner's share of investment in the company

d. General partnership
- Common among architectural, engineering firms

- Each partner can be held responsible for all company debts
- All partners share in start-up costs, making it easier to get business off the ground

e. Limited liability partnership (LLP)

- Similar to the general partnership
- Limits potential loss by each partner to only as his investment
- General partners in a business with limited liability partners have unlimited liability but are needed to run the business.

f. Corporation

- One group owns the company but a smaller group manages the business.

SLIDE 1-11

- Owners buy shares of stock (a part or slice of the business).
  1) Value of shares increases or decreases according to levels of success (or lack of success) of the company.
- Most small corporations and many large ones are privately held.
  1) Stock owned by a select group of investors
  2) Managed by board of directors appointed by stockholders at annual meeting of stockholders
  3) Board of directors meets regularly to decide on policies, major operating procedures of company
  4) Board appoints a president to manage day-to-day operation of corporation (Figure 1-9)
- Liability
  1) No one has unlimited liability
  2) Owners can only lose as much money as they have invested in stock.
  3) Individuals responsible for obeying the law and not trying to falsely control the finances of even a large corporation
  4) Government regulations for corporations are stricter due to limited liability involved with stock ownership.

8. Building Codes

a. Most towns, cities and counties have building codes.

b. A building code is a set of regulations that ensure that all buildings in that area are safe.

Slide 1-12
Slide1-13
Slide1-14
Slide1-15
Slide1-16
Slide1-17
Slide1-18

c. The local building department enforces the local building codes.

d There are many codes besides the building codes that govern the safe construction of buildings.

9. Unions and Contractors Associations

a. Two categories of construction organizations are of particular importance to construction students: craft unions and contractors' associations.

- A craft union is an organization of workers in a particular building trade.
- Some contractors' associations are formed to represent only non-union contractors, a few represent only union contractors, and others represent both.

SLIDE 1-19

Summary

The construction industry offers a wide variety of employment opportunities and potential careers. The jobs available range from unskilled labor positions all the way to being a part-owner in a corporation. The jobs in between these two extremes include apprentices, journeymen, engineers, architects, contractors, and owners. The key at each level is to be a good, conscientious worker, willing to advance through the educational and experience levels to achieve higher pay and jobs you enjoy doing. Each step along the way results in greater knowledge, more job security, career advancement, and the opportunity to work in the field or in an office. Every individual in the construction process plays an important role in achieving the finished product. Someone must lay the bricks, someone must design the subdivision, someone else must ensure that everything meets code and will bear up under the weight of the weather and the roof trusses, and someone must approve all of this and put his or her seal on the drawings. Other individuals provide the corporate backing and ownership needs. The construction industry is a place where anyone who wants to work at it can be successful.

Chapter 2 Working in the Industry

Instructor's Note

In this column, instructors will find a variety of lecture tips to aid student comprehension and notes on when to use PowerPoint slides to teach concepts.

Appendix B lists corresponding numbered PowerPoint Presentation titles

SLIDE 2-1 OBJECTIVES

OBJECTIVES

- Gain the ability to discuss ethical issues in the workplace
- Ability to work as a participating member of a work team
- Be able to communicate effectively with fellow workers
- Understand and discuss the value of lifelong learning

Glossary of Terms

Body language—sending unwritten, unspoken messages by how we hold our bodies or the expressions on our faces.

Ethics—the discipline dealing with what is good and bad and with moral duty and obligation.

Values—what we believe is right and wrong.

SLIDE 2-2

Teaching Tip: A discussion on ethics should include on-the-job situations they may deal with.

A. Ethics

1. We must combine our values of right and wrong with action. To know what is right and not do it shows a lack of ethics.
2. Having a good work ethic is giving everything you've got to your job and always trying to give the very best effort you have.
3. Rationalization can cause us to make poor choices for what we rationalize to be good reasons or causes.
4. Those individuals who operate in an ethical manner tend to live more upbeat and cheerful lives. (Figure 2-1)
5. Individuals who bypass values and ethics tend to become unhappy workers who no longer give their best effort.
6. Employees who are seen as pleasant, cooperative, hard working, and honorable are employees who employers feel good about advancing.

SLIDE 2-3

Teaching Tip: A reference to a local sports team will help send this message home.

B. Working on a Team

1. Constructing a building involves cooperation among the carpenters, plumbers, electricians, painters, masons, and others. (Figure 2-2)
2. Working as a part of a team means realizing that everyone involved with the construction process is equally important. There are no solo acts in the construction industry, only team players.
3. Traits of effective teams
   a. Listening: Team members listen to others and build on teammates' ideas.
   b. Questioning: Team members ask sincere questions of each other.
   c. Respect: Team members realize that the opinions and ideas of others are worthwhile.

d. Helping: Team members help each other.
e. Sharing: Team members readily exchange ideas and experience.
f. Participation: Team members contribute ideas, discuss them, and actively participate in projects.

C. **Communication**

SLIDE 2-4

Teaching Tip: You might have students try to explain how to do something simple, like tie a pair of shoes, without using their hand to explain this process, to emphasize good communication skills.

1. Good communication skills are vital for success in any career.
2. Without good communication it is impossible for the team to work together.
3. Speaking is one of the five basic forms of communication.
   a. A good vocabulary is important in order to clearly express what you mean to say.
   b. Poor grammar and incorrect word usage can cause problems on the job site and send the message that you are uneducated.
   c. Communicating is critical to earning respect as a professional and in earning people's business.
   d. Good communicators look a person in the eye and watch to see that the message they have sent has been received and understood.
4. Listening is just as important as speaking.
   a. In a world flooded with cell phones, radios, televisions, and computers it can become a bad habit to only half listen to what someone is saying. Effective listening requires focus and your full attention.
   b. What a person is saying to you might involve your safety or save you time by explaining how to complete a task more efficiently.
   c. Good listening can provide you with clues that can help you in obtaining or bidding on more jobs.
5. Writing is more important than you might think in the construction industry.
   a. The writer must keep the potential reader in mind.
   b. Make sure what you write is clear and easily understood. You can't see the reader and ask if that person understands what you wrote.
   c. Good penmanship, grammar, and vocabulary help ensure accuracy.
   d. Check spelling with a dictionary to ensure the message you wrote is what you intended.
6. Body language sends out as many messages as the spoken word does.
   a. What we do with our hands, posture, face, and eyes may send messages that we may not be aware of. (Figure 2-3)
   b. Always look people in the eye.
      - Looking around makes you seem uninterested.
      - Looking down gives the impression that you are untrustworthy.
   c. Your hands should be still.
      - Don't put them in your pockets.
      - Don't fidget.
   d. Appearance
      - Dress neatly and appropriately.
      - Work clothes should be neat and clean.
   e. Speak clearly
      - Don't mutter.

- Don't force your listener to strain to hear you. It gives the impression that you don't know what to say or you don't have anything worthwhile to say.

D. Customer Service

1. Good customer service also includes providing a good value for your fee, being honest, communicating clearly, being cooperative, and looking to provide the best possible experience your customer can have in working with you.

E. Lifelong Learning

SLIDE 2-5

Teaching Tip: Refer back to any guest speakers you have had, or work-place examples you have mentioned.

1. In the past, apprenticeships began in early teens for men, and resulted in workers who spent their lives in their chosen trade.
2. As technology continues to advance it is important for a worker to stay current with the skills and knowledge necessary to do a good job and stay competitive in the industry.
3. A worker who does not keep up with the changes in how a company is managed is destined to fall behind.
4. A lack of updated skills and techniques will prevent a worker from moving up and result being less valuable to the company.
5. Individuals who embrace lifelong and ongoing learning will continue to grow, feel good about themselves, and advance throughout their careers.
6. Adults of all ages and work histories continue to go to trade schools, community colleges, and universities to improve their current careers or to start brand-new careers.
7. The ongoing quest for knowledge and self-improvement enhances people's value as employees and business owners.

Summary

SLIDE 2-6

Successful employees have strong values and act upon those values. This is what makes them ethical workers. Ethical workers are dependable, hard-working, and honorable employees who are assets to the company they work for. Good communication skills are key to making sure that the messages you send are received correctly and understood. These communications may be verbal, written, in body language, and even unspoken. Our actions also speak for us, making our work ethic a type of communication to our employers as to what type of employees we are. Communication is important for the team effort required to successfully complete a construction project. It is important to remember that there are no solo acts, only great team players. It takes electricians, builders, plumbers, and others to all work and communicate together in order to build what the blueprints call for. Finally, we are never too old to learn new things. No one ever knows everything there is to know about something, and as quickly as technology changes, no one can ever be totally up to date on the newest techniques. We must remain open and teachable, realizing that learning takes a lifetime.

Chapter 3 Job Safety

Instructor's Note

In this column, instructors will find a variety of lecture tips to aid student comprehension and notes on when to use PowerPoint slides to teach concepts.

Appendix B lists corresponding numbered PowerPoint Presentation titles

SLIDE 3-1
OBJECTIVES

OBJECTIVES

- Have a complete understanding of accidents and their causes
- Define OSHA and its impact on construction workers
- Be able to identify the appropriate PPE for common work situations and explain how to use it
- Explain the Hazard Communication Standard and be able to use a Material Safety Data Sheet
- Be able to explain how fires start, are sustained, and extinguished
- Understand trench safety procedures
- Have an understanding of electric shock and be able to list safety considerations when working with or near electricity

Glossary of Terms

Amperage—the unit of measurement of electrical current.

Class A fire—fires that involve ordinary materials such as paper, cardboard, and wood. Class A fires can be extinguished with water.

Class B fire—fires that involve flammable liquids. These fires can be put out with either dry chemicals or CO2 (carbon dioxide).

Class C fire—fires involving electricity. Class C fires are extinguished with CO2 (carbon dioxide).

Conductor—a material that allows electricity to flow.

Fire triangle—the elements of heat, fuel, and oxygen (the three sides of the triangle) must be present for a fire to burn.

GFCI—Ground Fault Circuit Interrupter: A protective device that opens the electrical circuit when an imbalance in the amount of current flow between conductors or a circuit is detected.

Ground (electrical)—a conducting body that serves as the common return path for an electrical circuit. A ground typically has zero potential. The earth may also be used as a ground.

Horseplay—practical jokes and playful activity that are totally inappropriate on a construction site.

Labels—product labels containing valuable information.

MSDS—Material Safety Data Sheet: An MSDS is required to be available for any substance that might be harmful. This sheet gives complete information about the product and what to do in the event of exposure.

OSHA—Occupational Safety and Health Administration at both the state and federal levels. Also, the Occupational Safety and Health Act, which is administered by the Administration. OSHA generally refers to the laws that are intended to keep workers safe.

PPE—personal protective equipment: Any safety equipment you wear to protect yourself from safety hazards.

Right to Know Rule—OSHA rule that every worker has a right to know about any substances on the job that might be harmful to humans.

Voltage—The unit of measurement used to represent the electrical pressure that causes current to flow. Voltage is sometimes called electromotive force (EMF) because it is the force that causes electrons to move. Voltage also represents the difference of potential or potential difference in a circuit.

Working Conditions—the things in the work environment that affect your work

Work Practices—the things a worker does and how he or she works has a lot to do with safety.

VIDEO #1
CHAPTER 3

SLIDE 3-2

Teaching Tip: It would be good to have some examples of work related accidents that you are familiar with. The more common and least expected ones are the best for making your point about job safety.

A. Accidents

1. Poor work practices cause many workplace accidents.
   a. Untrained workers who want to feel mature and show how capable they are try to do jobs they have not been trained to do endangering themselves and those around them.
   b. A failure to read labels and instructions relating to job materials, tools, and installation procedures can lead to accidents.
      - Use of flammable, caustic, or adhesive materials in nonventilated areas endangers everyone in that area.
   c. Ignoring safety devices
      - Can cause serious injury
      - Does not allow you to work faster or better according to experienced carpenters.
   d. Horseplay
      - Absolutely no place on a job site
      - Any number of ways a person can be injured
        1) Knocking into scaffolds
        2) Spilling caustic substances
        3) Causing someone to receive an electrical shock
        4) Knocking materials from overhead platforms
      - Take care around the people you must trust to be safe around you.
   e. Poor communications
      - Communications failure can result in severe injury
      - Clear, accurate communications are critical
      - Misunderstanding could lead to injury
   f. Alcohol, drugs have no place on a job site
      - Prescription drugs
        1) Impair judgment
        2) Slow down reflexes
        3) Cause drowsness
2. Working Conditions
   a. Poor housekeeping can result in hazards that can be stepped on or tripped over, resulting in an injury.
      - Tools
      - Scrap materials
      - Exposed nails

b Clean up at the end of your shift to ensure a safe working environment for everyone.

c. Wet conditions that result in puddles should be dealt with by bridging with planks.
   - Never use power tools around water.

d. Excessive heat or cold can affect your work.
   - Dress according to the conditions that you know you will have to work in each day.

e. Poor lighting can lead to problems.
   - Portable lighting ensures that you can see what you are working on and the area you are working in.

f. Poorly maintained equipment
   - One of most common causes of accidents
   - Double-check tools and machinery for abnomalities prior to using items.
     1) Cracks
     2) Tears
     3) Leakage

g. Fire, explosion hazards
   - Keep chemicals, substances that can cause fires away from each other.
   - Keep areas properly ventilated.
   - Be aware of the nearest fire extinguisher; know how to use it.

VIDEO #1
CHAPTER 3

SLIDE 3-3

B. OSHA

1. Background
   a. Congress passed Occupational Safety and Health Act in 1970.
   b. Requires employers to provide safe workplaces
      - Workplaces must be free of serious recognized hazards.
      - Workplaces must comply with occupational safety and health standards.

Teaching Tip: You might make reference to any OSHA investigated accidents in your area. Along the gulf coast there are refinery incidents, there have been several explosions related to fireworks warehouses, and there may have been some situations in your area that you could use.

2. Occupational Safety and Health Act
   a. Authorizes states to establish their own safety and health programs with OSHA approval
   b. Grants workers certain rights
   c. Workers should work with employers to help ensure there is a safe working environment.
   d. Workers required to comply with all safety and health standards that apply to their actions on the job.
      - Employees should read the OSHA poster that every employer is required to display (Figure 3-3).
      - Employees should follow the employer's safety and health rules.
      - Employees should wear and use all required gear and equipment.
      - Employees should follow safe work practices on the job, as directed by the employer.
      - Employees should report hazardous conditions to a supervisor or safety committee.
      - Employees should report hazardous conditions to OSHA if employers fail to address problems.
3. Employers' responsibilities

a. Required to provide a safe and healthful workplace free of recognized hazards and to follow OSHA standards.

b. Required to provide training, medical examinations and record keeping.

c. Required to supply a *competent person* to build certain safety devices or to inspect safety protection of workers.

4. OSHA standards and rules are to protect workers against hazards on the job.

   a. Limits placed on the number of hazards an employee can be exposed to.

   b. Failure to comply can result in citations and fines.

5. Personal Protection Equipment (PPE) is an important aspect of safety.

   a. Worn for protection against safety hazards.

   b. Some equipment is designed to protect against specific hazards

      - Safety harness (falls)
      - Helmet (head injury)

Teaching Tip: It would be good to have several items of safety equipment available for students to see and handle. We remember more of what we see, handle, read about AND hear about all at the same time.

   c. Safety shoes (Figure 3-4)

      - Available with hard toes to protect against crushing by heavy objects.
      - Punctureproof soles protect against exposed nails or other sharp objects.

   d. Hard hats (Figure 3-5)

      - Protect against injury from falling objects
      - Should not be made of material that conducts electricity

SLIDES 3-4 TO 3-9

   e. Safety glasses (Figure 3-6)

      - Protects against small particles injuring the eyes
      - Approved glasses stamped with Z87.1 and have side shields

   f. Particulate mask

      - Usually disposable
      - Worn over the nose and mouth
      - Protect against airborne particles of dust (Figure 3-7).

   g. Safety harness (Figure 3-8)

      - Worn whenever a fall of more than 6 feet is possible
      - Attached to a secure anchor point to prevent falling all the way to the ground

   h. Ear protection (Figure 3-9)

      - Required when working around loud equipment
      - Level of protection available varies
        1) Length of time of exposure
        2) Decibel level (loudness) of sound
      - Wide range of protection device types available
        1) Simple earplugs
        2) External muffler ear coverage

SLIDE 3-10

C. Hazard Communication Standard

1. Often referred to as the Right to Know Rule

   a. Worker has the right to know what potential or existing substances on the job site might be harmful to humans.

2. Many of materials used in construction are hazardous.

   a. Can cause sickness or even death

   b. All labels need to be read and all instructions carefully followed.

3. HazCon Standard or Worker's Right to Know
   a. Every worker who comes in contact with any potentially dangerous material must have access to a *Material Safety Data Sheet* (MSDS).
   b. The MSDS describes the risks, how to guard against them, and what to do in case of a hazmat exposure (Figure 3-10).
4. Simple rules to eliminate most of the danger from common hazardous materials

Teaching Tip: Try to have some common construction related chemicals, solvents and rags available to show what you are talking about

   a. Always keep chemicals and solvents in their original containers with the manufacturer's label in place.
   b. Read the manufacturer's label completely before using even the most ordinary supplies.
   c. Read the MSDS for each hazardous or questionable product with which you come in contact.
   d. Use chemicals and other supplies only for their intended purpose.
   e. Follow all recommended safety precautions.
   f. Use the recommended PPE for every product and operation.

SLIDE 3-11
SLIDE 3-12

D. Fire

1. The Fire Triangle
   a. Heat raises the fuel to its flash point, the temperature at which it ignites.
      - Heat sources
        1) Open flame
        2) Smoldering or hot substances near the fuel
        3) Electric spark
        4) Chemical reaction
      - Spontaneous combustion can occur when a chemical reaction takes place in a confined space where no ventilation is available to remove the heat given off by the chemical reaction.
      - A pile of oily or solvent-soaked rags can result in spontaneous combustion as the chemicals react with the fabric in the rags.
   b. Fuel for a fire can be anything that is capable of being oxidized rapidly enough to give off heat.
      - Wood
      - Paper
      - Rags
      - Plastics
      - Solvents
      - Paint
      - Electrical insulation
   c. Oxygen feeds a fire. Some welding processes use pure oxygen supplied in tanks. Use care around oxygen tanks.

SLIDE 3-13

2. The Fire Tetrahedron is a fourth element involved in a fire (Figure 3-12).
   a. Fourth element is the chain reaction.
   b. The chain reaction occurs when the fuel is broken down by the heat of the fire, making it easier to oxidize.
3. Extinguishing Fires
   a. Eliminating one or more parts of the Tetrahedron
   b. Not all fires are extinguished the same way.

SLIDE 3-14

- Class A fires involve ordinary materials like paper, cardboard, and wood and can be extinguished by water and some dry chemicals. The key is to eliminate the ability of the oxygen to feed the fire.
- Class B fires involve flammable liquids like paint, solvents, and gasoline and can be extinguished with either dry chemicals or CO_2 (carbon dioxide). Never use a water extinguisher on a class B fire.
- Class C fires are electrical fires, and before they can be extinguished the electrical supply should be disconnected. CO_2 extinguishers are used to extinguish class C fires.
- Class D fires involve combustible metals like magnesium. A class D fire extinguisher should be used with this type of fire.

Teaching Tip: Have an example of the various fire extinguishers available for students to see. The best scenario is for you to have a controlled fire that they can help put out. You may not be equipped for this type of hands-on experience, but if you can, all the better.

c. The label on a fire extinguisher will tell you for what type of fire that extinguisher is rated (Figure 3-13).

SLIDE 3-15

E. Trench Safety

1. OSHA definition
   a. An excavation that is deeper than it is wide
   b. Less than 15 feet wide at the bottom
2. Common at construction sites
   a. Water lines
   b. Gas lines
   c. Electrical lines
   d. Sewer lines
3. Soil weight
   a. 2,000 to 3,000 pounds per cubic yard
   b. Pressure or vibrations from machinery can cause soil to collapse.
4. OSHA rules for trench safety
   a. Slope of the trench walls
   b. Type of shoring
   c. Never work in a trench unless it has been deemed safe by a *competent person* according to OSHA guidelines.

VIDEO #1
CHAPTER 3

F. Material Handling

1. Always lift with your legs and not your back.
2. Do not lift heavy objects by yourself; have a partner help.

SLIDE 3-16
SLIDE 3-17
SLIDE 3-18

G. Electricity

1. Electrical fundamentals (Figure 3-14) can be discussed more easily by comparing electricity to water moving from a fountainhead through pipes to a specific point.
   a. The pump forces water through the pipes, out of the fountain head, and back into the reservoir much like a battery supplies the push—the voltage needed to send electricity on its way (Figure 3-15).
   b. The switch in Figure 3.15 turns the electricity on and off by breaking and completing the circuit.
   c. A short circuit is similar to a hole in a pipeline (Figure 3-16). It allows leakage along the pathway, which diverts electricity away from its intended destination (Figure 3-17).

VIDEO #1
CHAPTER 3

2. Electrical Safety is crucial on a job site; electrical shock is a serious injury.
   a. The severity of an electrical shock depends on a variety of factors.

- The size, weight and the physical makeup of a person
- How well a person is connected to the electrical source
- Where the electricity enters and exits the body
- The actual amount of electrical current (Figure 3-18)

Teaching Tip: Have some chords, plugs, and adapters on hand for students to view and handle.

b. It is never wise to try to bypass the grounding prong on a three-prong plug.

c. If a power tool is not properly grounded and a short occurs somewhere in the system, you may receive a fatal dose of electricity when you pick the tool up, providing the current with another path to ground (Figures 3-19 through 3-22).

3. Ground Fault Circuit Interrupter
   a. Detects imbalances between neutral conductor, hot conductor, and trip
   b. Removes power from the tool or machine where the short exists
4. Protecting against electric shock
   a. Best accomplished by not coming into contact with electricity
   b. Stay focused on the task.
   c. Use safe procedures.

SLIDE 3-19

- Know the location of the electrical panel or disconnect before beginning the work.
- Have all workers complete a CPR certification course.
- Be certain that all tools, test equipment, and power cords are in good condition and working order.
- Do not use electrical devices in wet or damp locations unless they are specifically designed for this environment.
- Use fiberglass ladders.
  1) Wood and aluminum ladders can conduct electricity if they come in contact with a live circuit.
- Do not work on energized circuits.
- Before performing any type of task, think about the consequences.

Teaching Tip: Go over these safety facts one at a time and consider testing students on just this information, to make sure they memorize them.

- Do not take chances or cut corners.
- Always work carefully.
- Be sure that you are working in the safest environment possible.
- Remove all jewelry; gold is an excellent conductor of electricity.
- Discard all worn, defective, or otherwise unsafe items.
- Deenergize, lock out, and tag-out all electrical circuits upon which work is to be performed.
  1) Follow all lock-out and tag-out procedures exactly.
  2) Do not skip any steps in this area. (Figures 3-27 and 3-28)

SLIDES 3-20 TO 3-23

- Verify that the circuit is deenergized before beginning your work.
  1) Perform a visual inspection of your meter and test leads.
  2) Determine the amount of voltage that you expect to measure.
  3) Verify that the voltage rating of your meter and test leads will not be exceeded.
  4) Set up your meter to measure voltage as follows:

     Insert the black test lead into the *COM* test lead jack.

     Make certain the test lead is fully and securely inserted into the *COM* jack.

5) Insert the red test lead into the *Volts* or *V* test lead jack.
6) Determine the type of voltage (AC or DC), and set the voltage selection switch (if so equipped) to the proper voltage type.
7) Determine the approximate amount of voltage to be measured and set the range switch to the proper voltage range.
8) If you are unsure of the amount of voltage that you will be measuring, *set the range switch to the highest voltage range.*
9) Double check your meter setup.
10) Connect the red and black test probes to a circuit with a similar known amount of voltage present.
11) Read the indicated voltage from the digital display of your meter.

    If the indicated voltage is correct you can move to the next step.

    If it is not correct, your meter is faulty and must be repaired.

    Do not proceed until you have done this procedure with a properly functioning meter.
12) Connect the red and black test probes to the circuit that you are troubleshooting.
13) Read the indicated voltage from the digital display of your meter. You should have a reading of zero volts on a deenergized circuit.
14) Carefully disconnect your meter from the circuit that you are testing.
15) Reconnect your meter to the circuit with the similar known amount of voltage present that you previously used.

Teaching Tip: With an electrical box, demonstrate proper lock-out/tag-out procedure.

SLIDE 3-24

Summary

Safety can never, ever for any reason be shortchanged. To cut corners in safety is to endanger your life and the lives of those around you. OSHA standards have been established to help both the employee and the employer take active roles in job-site safety. Always be conscious of what your job will be each day and whether or not you have the proper PPE for the tasks you will be required to complete. Stay focused on what you are doing, remembering to keep certain solvents and materials away from each other to avoid a fire or caustic hazard. Use your knowledge of how fires start to avoid causing a fire and to be prepared to help extinguish one. Never work in a trench unless it has been proclaimed safe by a "competent person" according to OSHA standards. Remember how little electricity it takes to injure or kill someone, and make sure you take extra care to follow all safety procedures when working with power tools or in an environment where electrical current is flowing. Never take safety for granted. Stay alert and be aware of what is going on around you.

Chapter 4 Safety with Ladders, Scaffolds, and Machines

Instructor's Note

In this column, instructors will find a variety of lecture tips to aid student comprehension and notes on when to use PowerPoint slides to teach concepts.

Appendix B lists corresponding numbered PowerPoint Presentation titles

OBJECTIVES

SLIDE 4-1
OBJECTIVES

- List safety considerations in erecting and working on scaffolds.
- Be able to choose the right ladder for a job and use it safely.
- List the safety considerations in working with internal combustion engines and engine-driven machines.
- List the safety considerations in working with compressed air and pneumatic tools.

Glossary of Terms

4:1 Rule for ladders—ladders should be 1 foot away from the vertical surface against which they are placed for every 4 feet in height.

Competent person—one who is capable of identifying existing and predictable hazards in the surrounding area, or working conditions that are unsanitary, hazardous, or dangerous to employees, and who has the authorization to take corrective measures to eliminate them.

Portable ladder—a ladder that can easily be picked up and carried to another location.

Pump jacks—a mechanism that attaches to a vertical pole and can be pumped up and down the pole. Pump jacks are used to support planks on which workers stand.

Scaffolds—temporary work platforms.

SLIDE 4-2
SLIDE 4-3

A. Scaffolds
   1. Temporary work platforms widely used throughout the construction industry.
   2. Fabricated frame scaffolds (Figure 4-1)
      a. Versatile, economical, easy to use
      b. Stackable for large-scale jobs
   3. Pump jack scaffolds (Figure 4-2)
      a. Second most popular type
      b. Used when the work needs to move frequently
      c. Can be raised and lowered quickly by the pump mechanism
   4. OSHA requirement
      a. All scaffolds must be built by a "competent person"
         - Done in an effort to reduce injuries
         - 4,500 injuries, 50 deaths occur every year in scaffold-related accidents.
   5. Certified competent person builds scaffold with all existing situations and conditions at a job site taken into consideration

Teaching Tip: Explain what a competent person is, by OSHA standards, and how someone receives that label.

6. Key elements for working safely on a scaffold
   a. A stable scaffold can only be built on a square and level foundation.
   b. The scaffold should be built to OSHA standards.
   c. The scaffold must be built on a firm base with appropriate base plates and sills.
   d. Scaffold poles, frames, and uprights must be plumb and braced to prevent swaying and displacement.
   e. Frames and panels must be connected by cross, horizontal, or diagonal braces, alone or in combination, that secure vertical members together laterally.
   f. All brace connections must be secured to prevent dislodging.
   g. Employees must have safe access to any level of a scaffold that is 2 feet above or below an access point.
   h. OSHA standards specifically forbid climbing cross-braces as a means of access.
   i. Portable, hook-on, and attachable ladders must be positioned so as not to tip the scaffold.
   j. Top hazard of scaffolds is workers falling off. Fall protection involves personal fall-arrest systems or guard rails, and must be included on any scaffold 10 or more feet off the ground.

Teaching Tip: Have some color-coded tags to show students.

7. Color-coded tags (Figure 4-3) may be used to indicate scaffold inspections and safety status.
   a. Green—inspection has taken place; scaffold is safe
   b. Yellow—scaffold can be used but it does not meet all safety requirements; can only be used with a fall-protection harness
   c. Red—scaffold is not safe for use

VIDEO #1 CHAPTER 4

SLIDE 4-4

SLIDES 4-5 TO 4-11

B. Portable Ladders

1. Step ladders and extension ladders
   a. Most common ladder types
   b. Easily moved from one location to another
2. Using the right ladder for the job is important in preventing accidents.
3. Fiberglass ladders
   a. Use if any possibility that you will be anywhere near electricity
   b. Metal and damp or oil-soaked wooden ladders conduct electricity.
4. Choosing the right ladder for the job is always a wise choice (Figure 4-4):
   a. Type IA: Extra Heavy-Duty, 300 lbs
   b. Type I: Heavy-Duty, 250 lbs
   c. Type II: Medium Duty. 225 lbs
   d. Type III: Light Duty, 200 lbs
5. Follow all safety rules with all types of ladders.

VIDEO #1 CHAPTER 4

SLIDE 4-12

C. Internal Combustion Engines

1. These engines are powered by both gasoline and diesel fuel; include generators, compressors, power trowels, etc. (Figure 4-5).
2. These engines present several safety concerns:
   a. Fire caused by fuel and exhaust
   b. Carbon monoxide poisoning from exhaust emission
   c. Powerful moving parts that can entangle clothing, hair, fingers, and hands

   d. Electrical shock from the ignition system
   e. Hearing loss caused by operation
3. Following sensible safety rules can help you avoid injuries.
   a. Do not operate engine-driven equipment until you have been instructed in its proper, safe use.
   b. Keep motor fuels in approved containers made for that type of fuel.
   c. Keep open flame and sources of sparks away from fuel.
   d. Do not refuel a hot engine.
   e. Do not use gasoline or diesel fuel as a solvent or cleaning agent.

Teaching Tip: Emphasize safety considerations and potential hazards over and over. Make sure students understand that these can be life and death issues.

   f. Have a fire extinguisher for class B fires on hand; know how to use it.
   g. Never operate an internal combustion engine indoors.
   h. Keep engine-driven equipment in good condition.
   i. Do not remove guards, shields, or other devices intended to protect the user from moving parts.
   j. Keep engines dry.
   k. Do not operate an internal combustion engine without a working muffler.
   l. Wear hearing protection if you must work near the engine.
   m. Remove the spark plug wire(s) before doing any repairs to the engine or attached machinery.

SLIDE 4-13

D. Compressed Air

1. Used to power tools such as pneumatic nailers at a pressure of 100 to 200 pounds per square inch (psi).
2. Compressed air is powerful and safety is a must around these devices.
   a. Wear safety equipment.
   b. Use tools only for the purposes you have been trained to use them for.

Teaching Tip: Any pneumatic tools you can bring for students to handle will help in understanding this section.

3. Most pneumatic tools have quick-disconnect fittings (Figure 4-6). Hoses without quick-disconnects must have the air turned off before disconnecting the tool.
4. Other gases such as oxygen and refrigerants are also stored under pressure in cylinders. Always be aware of what you are working with and use all safety procedures.
5. Never use a tool powered by compressed air until you have been thoroughly trained to do so.

SLIDE 4-14

Summary

Safety is one of the most important words construction workers need to know. The majority of injuries occurring on construction sites could have been prevented if proper safety measures had been followed at all times. Scaffolds must be built by a competent person who understands all the elements involved in erecting a safe device. Ladders must be used wisely. Choose the correct ladder for each job. Remember that if you are going to be anywhere near electricity, select a fiberglass ladder. Special care should be taken around engines and the mechanisms they give power to. Additional care should be used around tools powered by compressed air. At no time is there ever a place for anything less than complete focus on a job site. When safety is at the forefront of any work you do, then you will continue to do good work.

Chapter 5 Whole Numbers

Instructor's Note

In this column, instructors will find a variety of lecture tips to aid student comprehension and notes on when to use PowerPoint slides to teach concepts.

Appendix B lists corresponding numbered PowerPoint Presentation titles

SLIDE 5-1 OBJECTIVES

OBJECTIVES

- Be able to add, subtract, multiply, and divide whole numbers.
- Solve problems involving multiple operations with whole numbers.

Glossary of Terms

Difference—the results of subtraction.

Factor—a number that, when multiplied by another factor, produces another number.

Minuend—the number from which another number is to be subtracted

Subtrahend—a number that is to be subtracted from another number.

Sum—the result of addition.

Teaching Tip: If you have a chalkboard, dry-erase board, or overhead projector have students do practice math problems on the board. It is easier to correct problems as they are happening than returning graded papers a week later.

A. Basic Principals

1. In order to perform the necessary mathematical functions required on the job, you must memorize addition tables, multiplication tables, rules of order, etc.
2. You must be able to line up columns of numbers correctly.
3. When working with units of measure such as feet, pounds, and quarts be sure to label your answer with the units you are working with (example: 2′ + 4′ = 6′ or 2 lbs + 4 lbs = 6 lbs).

SLIDE 5-2

B. Addition Practice

1. Solving for molding needed in a room with four walls
2. Adding up feet in a house to solve for total length (Figure 5-1)
3. Figuring your time sheet
4. Using column addition to figure ductwork needs
5. Adding up current load using branch circuits (Figure 5-2)
6. Adding cubic yards of concrete
7. Adding different types of cable to get one cable sum
8. Using column math to figure BTUs
9. Using column math to determine total gallons of paint needed
10. Adding sides to find a perimeter (distance around an area)

SLIDE 5-3

C. Subtraction

1. Actual practice with problems, using units
2. Practical use of subtraction with sawing boards on the site
3. Using simple subtraction to figure tubing
4. Subtracting cubic yards

5. Looking at a floor plan to determine information in order to first add one set of numbers to be subtracted from the dimension of wall C (Figure 5-3)
6. Figuring distance by looking at a floor plan and being able to subtract an inner dimension line from an outer dimension line in order to find a called-for distance (Figure 5-3).
7. Being able first to read the complete dimension of one wall, and then using subtraction to fill in a missing dimension on a parallel wall of the same length (Figure 5-3).
8. Using addition and then subtraction to determine amount of existing supply of cable left over after using some.
9. Ability to read a diagram and use logical thinking to use the 6-foot given dimension that has been halved, to determine how much of the 18-foot dimension is included in dimension A (Figure 5-4).
10. Ability to read a diagram, add up existing measurements, and then subtract them from a larger amount to determine a mission dimension (Figure 5-5).
11. Using simple subtraction to determine a safety margin.

SLIDE 5-4

D. Multiplication

1. Practice basic skills making sure to carry the units being multiplied through to the answer.
2. Use simple multiplication to figure out an amount of door casings.
3. Use multiplication to figure out amount earned.
4. Use simple multiplication to determine amounts of various types of materials needed for specific jobs.
5. Use multiplication twice to determine an amount needed by first multiplying the first two numbers given and then multiplying that answer by the third piece of information given.
6. Use basic multiplication skills to price out a needed amount of materials.

SLIDE 5-5

E. Division

1. Practice using basic division skills and remember to include unit names of materials being worked with, such as gallons, pounds, etc.
2. Use basic division to determine how many pieces of material are needed for a particular job, such as replacing lights from a case of lights or installation of insulation per hour.
3. Using division to establish how many items are needed for hanging pipe (Figure 5-6), installing wall sockets (Figure 5-7), and installing light fixtures.
4. Figuring how much you are paid per hour can be done using division.

SLIDE 5-6

F. Combined Operations

1. The key to problems that ask for several types of answers is to read slowly and carefully.
2. Determine what information you are given and then decide what it is you need to provide the answer.
3. Looking at the problem given, decide what order of operations you should follow in order to arrive at the answer you need.
4. Use addition and then subtraction to determine first how much was used, and then to see how much you have left.
5. Use multiplication to figure out the first group of hours, then multiply the second set of figures and add your two totals together to determine a total number of hours or shingles.

6. Learn to add and subtract dimensions on a diagram to determine how many of a particular item are needed in order to ensure that what you are doing meets all codes.
7. It is important to be able to use common sense when mixing units in math. Always change feet to inches, pounds to ounces, etc., so you have common measurements to work with.
8. You will need to know how to use the types of measurements used for area, perimeter, radius, and circumference to figure how many nails, screws, or other items you may need to install a wall or a roof or some other structure.
9. Don't let long problems throw you.
   a. Take long problems one step at a time.
   b. Write down each section to keep it simple.
   c. Follow the math operations you know how to do.

SLIDE 5-7

Summary

The ability to do basic math on the job site is extremely important. A good construction worker will constantly be required to figure lengths, widths, and depths. Construction work involves being able to determine how many nails, boards, paint, workers, or other things are needed to accomplish the task. The cost of miscalculating the pitch of a roof or a sewer pipe or any number of other crucial measurements can be high. It is important to practice basic math skills, and make sure you can use one skill with another. Keep long, involved problems simple by breaking them down and dealing with them one step at a time. Practice reading diagrams and figuring lengths, widths, and heights of areas not specifically labeled. The more you work with numbers the more comfortable you will become when working with them on the job site.

Chapter 6 Decimals

Instructor's Note

In this column, instructors will find a variety of lecture tips to aid student comprehension and notes on when to use PowerPoint slides to teach concepts.

Appendix B lists corresponding numbered PowerPoint Presentation titles

OBJECTIVES

- Students will be able to add and subtract decimals.
- Students will be able to multiply and divide decimals.

Glossary of Terms

Decimal fraction—a number representing a quantity of less than 1 and expressed according to the decimal number system.

Decimal point—the dot used to indicate the separation between whole numbers and decimal fractions.

VIDEO #2
CHAPTER 6

SLIDE 6-1
OBJECTIVES
SLIDES 6-2 TO 6-4

A. Basic Principles

1. The decimal system is based on multiples of 10.
2. The values of the numerals are based on how far to the left or right of the decimal they are (Figure 6-1).
3. Decimal numbers can be added, subtracted, multiplied, and divided.
4. When working with decimals, it is important to keep track of the decimal point.
5. Decimal fractions are usually written with a zero to the left of the decimal point, but the zero does not affect the value of the number.

Teaching Tip: If you have a chalkboard, dry-erase board, or overhead projector have students do practice math problems on the board. It is easier to correct problems as they are happening than returning graded papers a week later.

B. Addition and Subtraction of Decimal Fractions

1. Adding and subtracting decimals requires lining up the decimals and then performing the mathematical operation, just as you do with whole numbers.
2. Keep track of the decimal point, lining it up in the problem, and placing it in the correct place in the answer.
3. In both addition and subtraction, the decimal point drops down in alignment with the numbers being added or subtracted.
4. Decimal addition is useful in measuring for many applications.
   a. Circuit resistance
   b. Shingles for a roof
   c. Pounds of refrigerant for an air conditioner
   d. Gallons of paint
   e. Material costs

SLIDE 6-5
SLIDE 6-6

C. Multiplication of Decimals

SLIDE 6-7

1. When multiplying decimals, decimal points do not have to be aligned.
2. Upon completion of the multiplication problem, count the number of digits to the right of the decimal in both of the numbers being multiplied; that is

how many spaces you should fill in to the right of the decimal in your answer.

3. In some cases, the answer will have a smaller number of digits than the number of decimal places required. In these cases, add as many zeros to the left of the product as is necessary to give the required number of decimal places (for example: 0.27 x 0.18 = 0.486, but you need to fill four spaces to the right of the decimal, so your answer will be 0.0486).
4. When an answer exceeds the number of decimal places required in the answer, the answer is rounded off to the highest number of places used in the information given in the problem (example: to round 0.63861 to three decimal places, increase the 8 to 9 and drop the 61, resulting in an answer of 0.639).

SLIDE 6-8
SLIDE 6-9

5. Being able to multiply decimals allows you to figure per hour, per foot, per gallon costs of materials and labor.
6. Being able to multiply and round off decimals is useful in working with wire, electrical problems, diameters of pipe, etc.

SLIDE 6-10
SLIDE 6-11

D. Division of Decimal Fractions

1. Dividing decimal fractions is done the same way that whole numbers are divided.
2. Move the decimal point of the divisor to the right as many places as is necessary to make it a whole number.
3. Move the decimal point of the dividend the same number of places to the right.
4. Add zeros to the dividend as necessary to allow you to move the decimal point.
5. Place the decimal point in the answer directly above the decimal point in the dividend.
6. Divide as if you are dividing whole numbers.
7. Division of decimals is useful when working with siding, flooring, wire resistance, costing out materials, duct work, etc.

Summary

SLIDE 6-12

Basic math principles are used when working with decimals. Decimals can be added, subtracted, multiplied, and divided just like whole numbers. When adding and subtracting decimals it is important to remember to align the decimal points and keep that alignment all the way through to the answer. When multiplying decimals, it is important to count the number of spaces to the right of the decimal in both numbers being multiplied and have the same number of spaces to the right of the decimal in your answer. In multiplication problems, to keep your answer to the same number of spaces required, you can round up or down. When dividing decimals, you move the decimal point of the divisor over until you are now dividing with a whole number. Move the decimal point of the dividend the same number of places to the right as you did for the divisor, adding zeros as is needed. Then, simply place the decimal point in your answer, directly above where the decimal point is in the dividend. Being able to add, subtract, multiply, and divide decimals is important in order to do all the measuring required at a construction site. It is a mental tool that is just as important as your hammer, your drill, or your tape measure.

Chapter 7 Fractions

Instructor's Note

In this column, instructors will find a variety of lecture tips to aid student comprehension and notes on when to use PowerPoint slides to teach concepts.

Appendix B lists corresponding numbered PowerPoint Presentation titles

OBJECTIVES

- Students will be able to add and subtract fractions.
- Students will be able to multiply and divide fractions.

Glossary of Terms

Denominator—the number on the bottom of a common fraction.

Fraction bar—the horizontal line in a common fraction.

Lowest common denominator (LCD)—the smallest denominator that can be used to express all of the common fractions in a set.

Mixed number—a number made up of a combination of a whole number plus a common fraction.

Numerator—the number on the top of a common fraction.

VIDEO #2
CHAPTER 7

SLIDE 7-1
OBJECTIVES
SLIDES 7-2 TO 7-3

Teaching Tip: If you have a chalkboard, dry-erase board, or overhead projector have students do practice math problems on the board. It is easier to correct problems as they are happening than returning graded papers a week later.

A. Basic Principles

1. A fraction is part of a whole.
2. An inch can be divided many ways; if it is divided into eighths, then any part of that measurement would be an eighth (Figure 7-1).
3. Equivalent fractions are fractions that have the same value.
4. A fraction's value does not change when the numerator and denominator are both multiplied or divided by the same number.
5. Sometimes it is necessary to find the lowest common denominator (LCD).
6. When giving an answer in fractions, that fraction should be written in its lowest terms (example: 2/4 should be written as 1/2).

VIDEO #2
CHAPTER 7

Teaching Tip: Use anything you can to demonstrate fractions. This could include pizza, a dozen donuts, or anything else that you can visually place into parts of the whole.

B. Adding Fractions

1. When adding fractions, express them with their LCDs, add the numerators, then write the sum of the numerators over the denominator.
2. Fractions can involve inches, gallons, hours, minutes, or anything that can be divided into parts.

C. Adding Mixed Numbers

SLIDE 7-4
SLIDE 7-5

1. Mixed numbers are whole numbers and a fraction, such as 6 1/2.
2. To change a mixed number to a fraction, multiply the whole number by the denominator of the fraction, add the numerator of the fraction to this amount, and place the final sum over the original denominator.

D. Subtracting Fractions

SLIDE 7-6
SLIDE 7-7

1. Fractions must have a common denominator in order to be subtracted.
2. When subtracting a fraction from a fraction, express the fractions as equivalent fractions having a common denominator.

3. Subtract the numerators, placing your answer over the original denominator.
4. When subtracting mixed numbers, follow the same process used in addition of mixed numbers to arrive at the LCD, and then subtract the numerators, writing the difference over your LCD.

VIDEO #2
CHAPTER 7

SLIDE 7-8

E. Multiplying Fractions

1. Multiply the numerators together and also multiply the denominators together. Write the product of the numerators above the fraction bar, and the product of the denominators below the fraction bar (1/2 x 4/5 = 4/10).
2. When multiplying any combination of fractions, mixed numbers, and whole numbers, write the mixed numbers as fractions.
3. Write the whole numbers as fractions with a denominator of 1 (the whole number 6 would be written as a fraction like this: 6/1).

SLIDE 7-9

F. Dividing Fractions

1. Division is the inverse of multiplication.
2. The inverse of 4/1 is 1/4.
3. To divide fractions invert the divisor and multiply.
4. Express the answer in lowest terms (example: 2/6 expressed in lowest terms is 1/3).
5. To divide any combination of fractions, mixed numbers, and whole numbers, write the number as a fraction.
6. Write the whole number over 1, invert the divisor, and multiply. Express the answer in lowest terms.
7. Division can be used to determine how many parts you can get out of a specified number of supplies, like a case of lights, a gallon of paint, or 5 pounds of refrigerant.
8. Division can also be used to determine an hourly wage or cost factor.

SLIDE 7-10

G. Decimal and Common Fraction Equivalents

1. A fraction is an expression of specific division, such as parts of an inch, a foot, a minute, a length of pipe, etc.
2. To change a common fraction to an equivalent decimal, divide the numerator by the denominator.
3. To express a decimal as a common fraction, write the number after the decimal point as the numerator of a common fraction. At the same time, write the denominator as 1 followed by as many zeros as there are digits to the right of the decimal point.

SLIDE 7-11

Summary

The construction site is a location where math is used all through the day. Knowing that a fraction is merely a way of expressing part of a whole item will help you determine what mathematical process you need to use to figure out your needs, or what you have left after using a portion of something. Fractions can be added, subtracted, multiplied, and divided. When dealing with mixed numbers, you should always change everything into fractions. Place the whole number over 1. With addition and subtraction, find the lowest common denominator before adding or subtracting the numerators. When multiplying, remember to multiply both the numerator and the denominator. Division is made easier by inverting the fractions and multiplying. Always express your answer in lowest terms. To express a decimal as a common fraction, write the number after the decimal point as the numerator of a common fraction.

Chapter 8 Linear Measure

Instructor's Note

In this column, instructors will find a variety of lecture tips to aid student comprehension and notes on when to use PowerPoint slides to teach concepts.

Appendix B lists corresponding numbered PowerPoint Presentation titles

SLIDE 8-1
OBJECTIVES

OBJECTIVES

- Students will be able to read a U.S. Customary scale accurately to within 1/16 inch.
- Students will be able to read a metric scale to the nearest millimeter.
- Students will be able to perform basic operations with linear measure.

Glossary of Terms

Metric system—a system of measurement based on 10 (often called the S.I. system).

Nominal dimensions—dimensions used to identify the sizes of many construction materials. These are the dimensions of lumber before it is dried and planed or the dimensions of masonry units including the mortar joints.

Perimeter—the distance around the outside of a shape.

Scale—the system of graduating a measuring device, such as a ruler into units and fractional parts of those units.

U.S. Customary measure—the system of measurements we use in the United States based on inches, feet, quarts, gallons, pounds, etc.

VIDEO #2
CHAPTER 8

SLIDE 8-2
SLIDE 8-3

Teaching Tip: Have a variety of rulers, tape measures and other measuring devices available for students to work with.

A. Reading a U.S. Customary Scale

1. The U.S. Customary Scale is what most measurements of distance in the construction industry are based on.
2. A construction worker must be able to read the foot and inch measurements on a tape measure.
3. The basic measurements are the yard (3 feet), the foot, and the inch.
4. An inch, if divided into two equal parts, has two halves of the inch.
5. The two halves of an inch can each be divided in half, making four equal quarters of an inch.
6. The four equal quarters of the inch can be further divided, resulting in eighths of an inch.
7. Eighths of an inch can each be halved, resulting in equal measurements of sixteenths of an inch (Figure 8-1).
8. The marks on a U.S. Customary scale are called graduations.
9. A scale that is graduated by sixteenths has five sizes of graduation marks, with the inch mark being the longest, and the marks becoming smaller in length, all the way down to the sixteenth marks (Figure 8-2 and 8-3).

SLIDE 8-4
SLIDE 8-5

B. Nominal Dimensions

1. These are the dimensions of a product before any allowances or adjustments have been made.

2. Lumber sizes, for example are in nominal dimensions. A 1 x 4 is actually a 3/4 x 3-1/2 ; a 1 x 6 is actually a 3/4 x 5-1/2; a 2 x 4 is actually a 1-1/2 x 3-1/2.

SLIDE 8-6

C. Combining Feet and Inches

1. When working with a combination of feet and inches it is best to turn the feet into inches so that you are working with one type of measurement.
2. Figure out how many feet you have and multiply that number by 12 (the number of inches in a foot) to determine how many inches you are working with. For example, if your measurement calls for a five-foot length, multiply 5 x 12 (5 feet times 12 inches) to get 60 inches.

Teaching Tip: Show several pieces of lumber that demonstrate nominal dimensions.

3. Once you know how many inches you have from the foot measurement add whatever loose inches you had. For example, if your original measurement called for 5 feet, 4 inches, then you take the 60 inches you got from multiplying the 5 feet by the 12 inches and add the 4 inches you already had, resulting in a total measurement of 64 inches.
4. This process works in reverse as well. If you have a measurement that reads 131 inches and you need to know how many whole feet you have, you simply divide by 12 and list your answer in whole feet, with any remainder still being listed as inches. For example, 131 divided by 12 equals 10 with a remainder of 11. So, your answer would be shown as 10 feet, 11 inches.

SLIDE 8-7

D. Metric System

1. The metric system is a decimal system based on dividing or multiplying by 10.
2. The basic unit of measurement in the metric system is the meter.
3. With the meter as its base, the metric system uses various prefixes to indicate units larger and smaller than a meter.
4. For units larger than a meter, the prefixes used are deka (10 meters), hecto (10 deka meters or 100 meters), and kilo (10 hectometers or 1,000 meters).
5. For units smaller than a meter, the prefixes used are deci (one tenth of a meter), centi (one one hundredth of a meter), and milli (one one thousandth of a meter).

SLIDE 8-8

E. Perimeter Measure

1. Measurement around the outside of something is called the perimeter.
2. To find the perimeter of a square or a rectangle, add the lengths of the four sides.
3. A rectangle has two equal sides, so to find the perimeter you multiply the length times 2 and the width times 2, and then add those two products together: $P = 2L + 2W$.

Teaching Tip: Put some actual examples of square or rectangular room or circular areas that students would encounter on the job site.

4. A square has four equal sides, so the perimeter is found by multiplying the length of one side times 4: $P = 4S$ (Figure 8-7).
5. The perimeter of a circle is called the circumference (Figure 8-8).
6. The circumference is determined by multiplying 3.1416 (π pronounced pie) times the diameter (distance across a circle) of the circle: $C = 2\,\pi D$ (Figure 8-8).

SLIDE 8-9

Summary

It is absolutely necessary to be able to work with measurements on a construction site. To build a house requires the ability to measure distances, lengths, widths, circumferences, and many other items. It is important to be able to read measurements accurately, regardless of what types of measurement indicators are being used. In America, the basic measurements you will work with involve the U.S. Customary Scale. Occasionally, though, you will also need to understand how to deal with the metric system. It is also important to be able to work with simple formulas to determine perimeter and circumference. Learning to work with measurements and formulas is one of the most important lessons a construction worker can master.

Chapter 9 Percent and Percentage

Instructor's Note

In this column, instructors will find a variety of lecture tips to aid student comprehension and notes on when to use PowerPoint slides to teach concepts.

Appendix B lists corresponding numbered PowerPoint Presentation titles

SLIDE 9-1
OBJECTIVES

OBJECTIVES

- Find any percentage of any given number.
- Determine the percentage that a part is of its whole.
- Calculate the total amount when given the principle, rate, and time.
- Solve basic problems involving simple interest.

Glossary of Terms

Amount—principal plus interest.

Compound interest—based on the principal plus previously earned interest.

Interest—payment for the use of money, expressed as a percent of the principal. Interest is usually stated.

Percent—hundredths of the whole.

Principal—the amount of money invested or borrowed.

Rate—percent at which interest is calculated.

SLIDE 9-2
SLIDE 9-4
SLIDE 9-5
SLIDE 9-8
SLIDE 9-9

A. Percent

1. Percent means each hundred, just as miles per hour means miles each hour.
2. The percent sign (%) may be considered a magnet that attracts the decimal point two places to the right.
3. The basic formula for percent: part (P) = percent (%) × whole (W).
4. To find percentage: % = P/W

VIDEO #2
CHAPTER 9

B. Percentage

1. The whole is sometimes called the base.
2. The percent is called rate.
3. The part is called the percentage.

C. Interest

1. Interest is a fee charged by a lender for the use of borrowed money.
2. Banks pay interest to depositors for the privilege of using the depositors' money.
3. Interest is usually expressed as a percent of the principal of amount borrowed.

SLIDE 9-10

Summary

It is necessary to be able to work with percentages at a construction site. Building a house requires the ability to determine percentage of materials used, purchased, and so on. You will need to be able to calculate percentage rates.

Chapter 10 Area and Volume

Instructor's Note

In this column, instructors will find a variety of lecture tips to aid student comprehension and notes on when to use PowerPoint slides to teach concepts.

Appendix B lists corresponding numbered PowerPoint Presentation titles

SLIDE 10-1 OBJECTIVES

OBJECTIVES

- Students will be able to determine the area of squares and rectangles.
- Students will be able to determine the area of triangles.
- Students will be able to determine the area of circles.
- Students will be able to determine the volume of rectangular solids and cubes.

Glossary of Terms

Area—the space inside a shape.

Base of a triangle—the side opposite the corner from which the height is measured. This can be any side of the triangle.

Circle—a shape in which every point on the perimeter is the same distance from a center point.

Cube—a three dimensional shape in which height, width, and depth are all equal.

Height of a triangle—the length of a line drawn perpendicular to one side of a triangle and extending to the opposite corner.

Rectangle—a shape with four 90° corners and with opposite sides of equal length.

Solid—a three dimensional shape. Spheres and cubes are two types of solids.

Square—a shape with four 90° corners and four equal-length sides. Square also refers to the result of multiplying a number by itself.

Triangle—a shape formed by three sides.

Volume—the space enclosed by a three dimensional figure.

SLIDES 10-2 TO 10-5

Teaching Tip: If you have some sample blue prints you can pass around, showing examples of these shapes, it would help students to see the on-site application of what they are learning.

A. Area of Squares and Rectangles

1. Always measured in units of square inches, square feet, square yards
2. One inch times one inch equals one square inch, just as one foot times one foot equals one square foot (Figure 10-1).
3. To find the area of a shape, remember to work with all the same units of measurement. You cannot multiply feet times inches or yards times meters.
4. In the U.S. Customary system, area is measured in square inches, feet, yards (Figure 10-2)
5. In the metric system the base unit of measurement is the square meter.
6. Often, you will find combinations of shapes that you need to find the measurements of. Measure each shape independently, and then add those results together (Figure 10-3).

SLIDE 10-6

B. Areas of Triangles

1. The area of a triangle is solved for by multiplying 1/2 x the length of the base x the height (Figure 10-8).

2. Remember to convert feet to inches and to work with only one unit of measurement.

C. Areas of Circles

SLIDE 10-7

1. To determine the area of a circle you will use a constant number called π (pronounced pie) that equals 3.1416 or 22/7.
2. The formula also includes the radius of the circle, which is half of its diameter.
3. The formula for measuring the area of a circle is $A = \pi r^2$.

D. Volume of Rectangular Solids and Cubes

SLIDE 10-8

1. Volume is the space enclosed by a three-dimensional figure.
2. The volume of a rectangular solid is found by multiplying the area of the base by the height (Figure 10-12).
3. A cube is a solid in which the width, length, and height are all equal.
4. Volume is always expressed as cubic inches, feet, yards, etc.

Summary

SLIDE 10-9

A typical construction site may have several different shapes that have to be constructed. Roofs involve triangles, rooms can be squares, rectangles, or a combination of shapes. A spa could include a circular hot tub that will also require you to figure out the volume of water to be held in the tub. For these reasons, and many more potential challenges on the job site, it is important to be able to solve for the area of squares, rectangles, triangles, and circles. It will be equally necessary to be able to determine the volume of rectangular solids and cubes.

Chapter 11 Right Angles

Instructor's Note

In this column, instructors will find a variety of lecture tips to aid student comprehension and notes on when to use PowerPoint slides to teach concepts.

Appendix B lists corresponding numbered PowerPoint Presentation titles

OBJECTIVES

- Students will be able to determine the hypotenuse of a right triangle if given the two adjacent sides.
- Students will be able to determine the length of one side of a right triangle if given the other side and the hypotenuse.

SLIDE 11-1 OBJECTIVES

Glossary of Terms

Hypotenuse—the side of a right triangle that is opposite the right angle.

Pythagorean theorem—a mathematical law that says the sum of the squares of the sides of a right triangle are equal to the square of the hypotenuse.

Right angle—a 90° angle.

Right triangle—a triangle with a 90° angle.

Sides of a right triangle—the two sides next to the right angle. The hypotenuse is not referred to as a side.

SLIDES 11-2 TO 11-4

Teaching Tip: You may need to review high school math at this point. In some cases, this may be brand new material, so you will have to go slow and check for understanding many times during this section of the text.

A. Basic Principles

1. The Pythagorean theorem states that the sum of the squares of the sides of a right triangle is equal to the square of the hypotenuse.
2. The hypotenuse is the longest side of a right triangle and is located opposite the right or 90° angle (Figure 11-1).
3. Uppercase letters are used to represent angles when labeling triangles.
4. Lowercase letters are used to represent the sides opposite those angles.
5. In right triangles, those with a 90° angle, the letter C is used to represent the right angle (Figure 11-1).
6. Formulas are created by using the letter c to represent the length of the hypotenuse and the letters a and b to represent the two sides.
7. The Pythagorean theorem can be rearranged using these three letters: $c^2 = a^2 + b^2$ or $a^2 = c^2 - b^2$ or $b^2 = c^2 - a^2$.

SLIDE 11-5
SLIDE 11-6

B. 6-8-10 Method

1. This is a method used to check the squareness of corners.
2. The 6-8-10 method is an application of the Pythagorean theorem.
3. To use the 6-8-10 method, measure 6 feet along one line and 8 feet along another line. When those two points are exactly 10 feet apart, the corner is square (Figure 11-7).

SLIDE 11-7

Summary

Right triangles will quickly become a construction worker's friends. The math used to figure right triangles is the easiest way to solve layout and electrical problems. This same basic math is also adaptable as the 6-8-10 method. The 6-8-10 method is used to test the squareness of things on the job site that are supposed to be square. Practicing the use of these various formulas will allow you to add yet another mental tool to your construction tool kit.

Chapter 12 Combined Operations

Instructor's Note

In this column, instructors will find a variety of lecture tips to aid student comprehension and notes on when to use PowerPoint slides to teach concepts.

Appendix B lists corresponding numbered PowerPoint Presentation titles

OBJECTIVES

- Solve construction-related math problems using addition, subtraction, multiplication, and division.
- Solve construction-related math problems using the formulas for area and volume.
- Solve construction-related math problems while demonstrating the ability to read measuring devices.
- Solve construction-related math problems while demonstrating the ability to use fractions and percentages.

SLIDE 12-1
OBJECTIVES
SLIDE 12-2

Teaching Tip: Continue to double check for understanding and have students do as many problems as possible on the board. Use real-life construction site situations for students to solve for. Utilize all drawings in the text for this purpose.

A. Basic Principles

1. Always break down combination problems into the individual math steps that need to be applied.
2. Analyze each problem to determine what information is given and what you need to calculate.
3. Write down each step that you need to go through to find your answer or answers.
4. Remember to get each measurement into one type of measuring unit, such as converting feet to inches, for the calculation process.
5. Remember to use the formulas for figuring for the area of a square, triangle, and circle.
6. Remember to use the formulas needed to solve for volume when the situation calls for it.

SLIDE 12-3

Summary

Everything you learn in your construction classes is geared to make you a better, more efficient, and safer worker. By applying the lessons you've learned in earlier chapters, you become more and more proficient on the job site. This makes you a more valuable employee. Proper use of the various math calculations saves the company money and allows you to complete your job in a shorter amount of time without endangering anyone on the site. It is important to continue to use these skills and your growing ability in the field to determine the sizes of rooms, the pitch of a roof, the number of gallons of paint needed, how many hours it will take to complete a job, and many other things you will be required to do.

Chapter 13 Hand Tools—Selection, Use, and Care

Instructor's Note

In this column, instructors will find a variety of lecture tips to aid student comprehension and notes on when to use PowerPoint slides to teach concepts.

Appendix B lists corresponding numbered PowerPoint Presentation titles

SLIDE 13-1
OBJECTIVES

OBJECTIVES

- Students will be able to read a ruler or measuring tape to 1/16 inch.
- Students will be able to identify common tools and explain their use.
- Students will learn how to safely perform simple operations with common tools.

Glossary of Terms

Blade (rafter square)—the longer arm of a square.

Kerf—the cut made by a saw.

Level—parallel to the earth's surface.

Pitch (saw)—the coarseness of the teeth of a saw. Pitch is measured in points per inch.

Plumb—perfectly perpendicular to the earth's surface.

Target rod—a graduated pole used with a builder's level to measure elevation.

Tongue (rafter square)—the shorter arm of a square.

Tripod—three-legged stand for holding a builder's level or a transit.

SLIDE 13-2

Teaching Tip: It would be great to have these measuring devices for students to handle and become accustomed to.

A. Systems of Measurement

1. English system
    a. Uses inches, feet, yards to measure distances in construction
    b. Yards can be divided into feet and feet into inches
    c. Inches can be divided as far as the job requires (Figure 13-1).
2. Metric system
    a. Based on multiples of 10

VIDEO #3
CHAPTER 13

SLIDE 13-3

B. Meausring Tools

1. Tape Measure
    a. Steel tape measures (Figure 13-2)
        - Come in lengths ranging from 6 feet to 100 feet
        - 20- to 30-foot lengths used most often for measuring building parts
        - Shorter lengths usually have a sliding hook to allow both inside and outside measurements to be taken at the same time (Figure 13-3 and Figure 13-4).
        - Longer tapes have a fitting that can be slipped over a nail or hooked over an outside corner (Figure 13-5).
    b. Inspection and defects
        - Rewind carefully to keep tape measures from kinking.
        - Keep tape clean, dry

- Dirt, moisture cause markings on the face to wear away more quickly.
- Inspect fitting on the end of the tape to make sure it is not broken and works properly.

2. Rafter Square

SLIDE 13-4

   a. Also called a framing square
   b. Used for laying out or checking square corners (Figure 13-6).
   c. Shorter part of the square is called the tongue; longer part is the blade.
   d. Each edge of the rafter square is graduated with fractions of an inch.
   e. Surfaces have tables for laying out rafters and braces.
   f. Inspections and defects
      - Bent squares are useless; care needs to be taken not to drop them.
      - Rafter squares are tough; if proper care is taken they will last practically forever.

3. Stair Gauges

SLIDE 13-5

   a. Can be clamped to the edge of the rafter square (Figure 13-7).
   b. Used when several parts must be laid out with the same angles and dimensions.

4. Speed Square

SLIDE 13-6

   a. Triangular metal tool (Figure 13-8) with several scales used for marking the cuts most often made on rafters.
   b. An adjustable arm can be set at any angle.
   c. Convenient, easy to carry, quick to position and use
   d. Inspection and defects
      - Should not be bent or have nicks along the edge.
      - Dropping the square or using it inappropriately will damage it.

5. Combination Square

SLIDE 13-7

   a. This tool (Figure 13-9) has a movable head on a 12-inch blade.
      - Right-angle surface
      - 45°-angle surface
      - Small level
   b. Smaller than framing square but not as tough or as easy to carry as a speed square
   c. Inspection and defects
      - Blade, head, and awl should be in place
      - If the awl is missing, the rest of the square will work fine.
      - Dropping or using inappropriately will cause damage to the square.

6. Chalk Line Reel

SLIDE 13-8

   a. Used to mark long straight lines, such as needed for laying out a wall on a floor. (Figure 13-10)
   b. Reel case contains powdered chalk that coats the line as it is pulled from the reel.
   c. The chalk-covered line (Figure 13-11) is pulled tight and released, leaving a chalk line on the surface being marked.
   d. Masons use a similar line to level and lay bricks (Figure 13-12).
   e. Inspection and defects
      - Keep the chalk line clean, dry

- Dirt, water will foul the line and contaminate the chalk.
- Double-check to be sure you have chalk and that line is in good shape.

SLIDE 13-9
SLIDE 13-10

Teaching Tip: Work a few problems using the 6-8-10 method, making sure students understand how to do it, and why they would use this method.

7. 6-8-10 Method and Checking Diagonals
   a. Although not tools, these techniques are valuable for checking the squareness of large corners and rectangles.
   b. A triangle with sides of 6, 8, and 10 units of length contains a right triangle (square corner).
   c. If the corner is a perfect 90° angle, the distance between these points is exactly 10 feet (Figure 13-13).
   d. The squareness of a square or rectangle can also be checked by measuring its diagonals. When all four corners are 90°, the two diagonals are equal in length (Figure 13-14).

SLIDE 13-11

8. Spirit Level
   a. Tool that has one or more small transparent vials filled with mineral spririts for determining levelness or plumbness (Figure 13-15).
      - Level refers to being parallel to the ground.
      - Plumb refers to being perpendicular to the ground (Figure 13-16).
   b. Sizes range from very small to 8 feet in length.
      - Common sizes: 9 inches, 2 feet, 4 feet

Teaching Tip: Any levels you can bring to class will certainly help students understand how to use a level better. There are numerous things in a classroom that can be checked for levelness. Remember, the more hands-on experience you give your students, the better they will transfer that knowledge to the job site.

   c. Very small line-level (Figure 13-17) is among the many special types of spirit levels.
   d. Inspection and defects
      - Take extra care to avoid jarring or dropping the level; misalignment of the vial can occur, leaving the level useless.
      - Check for visible signs of damage prior to using the level.
      - Check it against another level that you know is in good condition if there is doubt as to the condition of the level.

SLIDE 13-12

9. Builder's Level
   a. Used for measuring vertical distances over large horizontal areas, such as corners of house foundations (Figure 13-18).
   b. Each part has a separate function.
      - Telescope
        1) Lens
        2) Focusing adjustment
        3) Crosshairs for sighting
      - Telescope Level has a spirit level used for leveling the instrument when it is set up for use.
      - Clamp Screw locks the instrument in position so that it cannot be turned off the target.
      - Fine Adjusting Screw makes fine adjustments to the left or right.
      - Leveling Base holds four leveling screws for leveling the instrument prior to use.
      - Protractor is a scale graduated in degrees and minutes (a minute is 1/60th of a degree) used for measuring horizontal angles.

SLIDE 13-13

   c. Builder's level should not be confused with a transit that can be tilted to measure vertical angles (Figure 13-19); the builder's level does not tilt.
   d. A tripod and a target rod are needed in order to use the builder's level (Figure 13-20).

SLIDE 13-14

Teaching Tip: It would be good for students to pair up and practice using a builder's level and target rod for measuring something like points on the parking lot. These measurements could then be used to practice other things students have already learned.

10. Measuring Elevations
    a. Set up your tripod
        - Try to be sure surface is as level as possible.
        - If on a flat surface, connect tripod legs with a chain so they don't spread out any farther than you want.
    b. Keep the level in its case until you are ready to use it.
    c. Carefully tighten all necessary clamp screws to make sure everything is tightly secured for accurate measurements.
    d. Be careful not to touch the tripod with your foot; the slightest jar of the tripod will cause the instrument to be knocked out of level.
    e. While a partner holds the target rod, sight through the telescope and focus the crosshairs on the target rod.
    f. Write down the measurement seen through the telescope.
    g. Your partner will move the target rod to the other end of what you are measuring while you carefully rotate the instrument without touching the transit.
    h. Focus on the target rod; write this measurement down.
    i. Subtract the smaller reading from the larger one to determine the difference in the elevation that you are measuring.

SLIDE 13-15
SLIDE 13-16

11. Laser Level
    a. Uses laser light
        - Can be seen as a small dot over great distances
        - More reliable than using eyesight through telescope
    b. Some laser levels are self-leveling; others need a tripod, much like the builder's laser (Figure 13-21).
    c. In measuring elevations, once the instrument is leveled the laser is turned on.
        - Puts out narrow beam of rotating light
        - Light strikes everything in its path at the same elevation (Figure 13-22)
    d. A receiver that can see the laser light in bright sunlight is attached to the target rod (Figure 13-23).
    e. All laser instruments are required to have warning labels attached (Figure 13-24).
    f. Only trained workers should set up and use laser instruments.
    g. Safety rules are to be followed.
        - Never look directly into a laser beam.
        - Never view a laser beam with an optical instrument (builder's level).
        - Set up the laser level so it is above or below eye level when possible.
        - Turn the laser off when not in use.
        - Do not point a laser beam at another person.

SLIDE 13-17

Teaching Tip: Supplying a plumb bob for students to see will be helpful here.

12. Plumb Bob
    a. A plumb bob (Figure 13-25) is a pointed weight that can be attached to a string.
        - Made of brass or steel
        - Gravity causes it to hang perfectly vertical or perpendicular to earth's surface.
    b. A perfectly vertical line is said to be plumb.

c. The plumb bob can be useful in determining where to place a support beam or line up a tripod over a precise point.

SLIDE 13-18

13. Hammers
    a. Claw hammers come in two types and varying weights. (Figure 13-26)
        - Curved claw
        - Straight claw
        - 16-ounce
        - 20-ounce

Teaching Tip: Bring a variety of hammers to class so students can see them as you discuss the use of each one. Remember to cover inspections and safety.

    b. Always check hammer for signs of cracks or splintered wood.
    c. Bricklayer's hammer
        - Also called mason's hammer
        - A square and a chisel-like cutting edge (Figure 13-27)
        - Used to split bricks
    d. Sledge hammer
        - Used for driving stakes and breaking up hard materials
        - Comes in various sizes
    e. Sledge hammer safety
        - Caution: a sledge hammer can cause serious injury.
        - Make sure that the head is secure.
        - Make sure the handle has no splits.
        - Never hold an item to be struck by a sledge hammer.
        - Always wear safety gear.

SLIDE 13-19

14. Bars and Nail Pullers
    a. A variety of pry bars are used in construction that all have notches in the ends that can be used for pulling nails. (Figure 13-29)
        - Wrecking bars
        - Flat bars
        - Cats paws

Teaching Tip: Bringing tools to class applies to any and all tools discussed in this text. Always remember to cover what they should be looking for in a visual inspection and reinforce safety rules.

    b. A bar allows you to apply much greater force than a claw hammer.
    c. Any sharp edges found during a careful inspection should be removed with a grinder prior to using the tool.

SLIDE 13-20

15. Screwdrivers
    a. Screwdrivers come with a variety of heads.
        - Slotted
        - Phillips
        - Torx
        - Square tips (Figure 13-30)
    b. Screwdrivers come in different sizes. Make sure you use the right size for the job (Figure 13-31).
    c. A common mistake is using a screw driver with a damaged head.
        - Always check a screwdriver for damage to the head and handle.
        - Damage often occurs when the screwdriver is struck by a hammer or turned using pliers.

SLIDE 13-21

16. Pliers
    a. There are many types of pliers (Figure 13-32) available in several sizes.
    b. Needle-nose pliers are used for small delicate objects.

c. Electrician's pliers have a cutting section for cutting wires and a gripping section.
d. Slip-joint or channel-lock pliers can be adjusted for the widest range of sizes.
e. Vise-grip pliers are used when maximum strength is needed.

SLIDE 13-22

17. Wrenches
   a. Socket wrenches
      - Consist of a socket and handle (Figure 13-33)
      - Available in sizes ranging from 1/4 inch to 2 inches and from 3 mm to 50 mm
      - Most common sizes on construction sites are 3/8 inch and 7/8 inch
      - Two basic handle types
        1) Flex handles where maximum force is needed
        2) Ratchet handles allow user to move the handle back and forth without having to remove the socket from the nut.
   b. Open-end wrenches
      - Have different sizes at each end (Figure 13-34)
      - Some are the same size on both ends but with each end angled differently.
      - Find the right wrench for the job so the nut or bolt being loosened or tightened is not damaged.
   c. Box-end wrenches
      - Different size at each end (Figure 13-35)
      - Ends form complete circles with either 6 or 12 points inside the circle for gripping.
   d. Nut drivers
      - Look like a cross between a screwdriver and a socket wrench (Figure 13-36)
      - Available for use with smaller bolts
      - Not made for applying a great deal of torque
   e. Adjustable wrenches
      - Have a fixed jaw and a moveable jaw (Figure 13-37)
      - Adjustment is made by turning the adjusting screw
      - Range in size from 4 inches to 2 feet; 10-inch and 12-inch wrenches are the most popular on a job site
   f. Pipe wrench
      - Another type of adjustable wrench (Figure 13-38)
      - Jaws have teeth
      - Force should be applied toward direction of moveable jaw
   g. Basin wrench
      - (Figure 13-39) Used to reach the large retaining nuts holding faucets in place on sinks
      - Self-adjusting wrench on a long tee handle
      - Wrench end pivots for installation or removal
      - Can have a telescoping handle that can be extended
   h. Inspection and defects
      - Using the wrench as a hammer can damage the metal and might damage the gripping surface.

- Check for cracks, burred edges, and stretched jaws, especially if the wrench has been hammered on.
- Damage can occur to the gripping surface if it has been used to loosen or tighten a nut or bolt that it did not actually fit.

SLIDE 13-23

18. Hacksaw
    a. Consists of a metal frame to hold a hacksaw blade and a handle (Figure 13-40)
    b. Blades are attached, teeth aimed away from the handle, and made taut by turning the adjusting screw or wing nut.
    c. Cutting is done with the forward stroke of the blade; the blade should be lifted slightly on the backstroke.
    d. Inspection and defects
       - Make sure there are no teeth missing and that the blade is sharp.
       - Make sure the handle is in good working order, tightly holding the blade in place.

SLIDE 13-24

19. Handsaws
    a. Used for sawing wood; different teeth arrangements are used for different types of sawing.
    b. The coarseness or pitch of a saw is measured in points per inch (Figure 13-41).
    c. The cut made by a saw, called a kerf, is usually about 1/8-inch wide.
    d. Crosscut saws (Figure 13-42) are used to saw wood across the grain.
    e. Rip saws are used to cut with the grain of the wood (Figure 13-43).
    f. Inspection and defects
       - Make sure that teeth are sharp and uniform.
       - Check for cracks in the handle.
       - Make sure the saw blade is tightly connected to the handle.
    g. Using a handsaw

SLIDE 13-25

       - Place the wood to be sawed on a stable rest, like two sawhorses; make sure the saw will not be cutting into anything else.
       - Use a square or other straight edge to mark where the cut is to be made.
       - Hold the saw in your hand (right hand if you are right-handed, left hand if you are left-handed) in a loose but controlled grip.
       - Position yourself in a comfortable stance out of the way of the saw.
       - Position the saw blade on the scrap side of the pencil line, allowing for the kerf the saw will make.
       - Hold the saw at about a 45° angle, and push the saw forward, allowing the weight of the saw to push the teeth against the wood.
       - The cut occurs on the forward stroke.
       - Apply very slight twisting pressure as necessary to keep the kerf on the scrap side of the wood.
       - Support the scrap side of the wood so that it doesn't break off, leaving a splintered end on the wood you need to use.

Teaching Tip: Alert! Remember to caution students against pulling a utility knife towards their bodies, and to be aware of where their hands are during all cutting and sawing procedures.

    h. Coping saws
       - Used by carpenters to cut the profile of one piece of molding on the end of another piece when fitting molding inside corners.
       - A coping saw (Figure 13-44) has a thin frame holding a crosscut blade.
       - This saw has the teeth pointing toward the handle, and the cut occurs

as the blade is pulled toward you.

i. Wallboard saws
   - (Figure 13-45) Used for cutting small openings in wallboard.
   - Coarse teeth allow quick cutting of gypsum wallboard quickly
   - Thin blades allow cutting of sharp curves

SLIDE 13-26

20. Utility knife
    a. Variety of uses (Figure 13-46)
       - Scoring gypsum wallboard
       - Cutting the backing on insulation
       - Cutting shingles
    b. Available in many styles
    c. Most have spare blades stored in the handle.
    d. Always use a sharp blade.
    e. Never cut toward yourself.
21. Snips
    a. Aviation snips (Figure 13-47) are used to cut thin sheet metal.
    b. Snips should not be forced into heavy metal or used to cut wire.

SLIDE 13-27

22. Pipe and Tubing Cutters
    a. Tube cutters (Figure 13-48) are used to cut aluminum, copper, brass, and steel tubing and plastic pipe.
    b. Tube cutters have two rollers in the lower jaw and a round cutter in the upper jaw.
    c. The blades are interchangeable; certain blades used for certain materials.
    d. Pipe cutters are larger versions of the same tool.
    e. The triangular piece of metal on the back of the tubing cutter is a reamer; it is used to ream or smooth the inside edge of the cut end of the tubing.
    f. Do not use a plastic pipe cutter (Figure 13-49) to cut anything other than plastic. To do so will damage the tool.
    g. Check for nicks and dullness on the blades
    h. Make sure that both rollers roll smoothly and that the cutting wheel is not loose on its axis prior to each use.

SLIDE 13-28

23. Flaring Tool
    a. The flaring tool (Figure 13-51) consists of a pair of flaring bars that, when clamped together, create holes the right size for various sizes of tubing.
    b. The flaring tool has a yoke that clamps around the flaring bars and the feed screw.
    c. The bottom of the feed screw is fitted with a cone-shaped end that matches the angle of the chambers in the flaring bars (45°).
    d. Operating instructions
       - Cut and ream the tubing.
       - Back off the feed screw and loosen the yoke to permit the flaring bars to slide freely through the yoke.
       - Insert the tubing into the proper size opening and close the flaring bars.
       - Push the tubing up from the bottom until it is just even with the top of the flaring bars.

- Slide the yoke into place over the tube.
- Tighten the feed screw until the tubing is flared against the sides of the flaring bar.
- Back off the feed screw as far as it will go. Loosen the yoke and slide it back so the flaring bars can be opened.

SLIDE 13-29

24. Swaging Tools
    a. Swaging is a process used to allow one piece of tubing to fit inside the enlarged end of another piece of tubing.
    b. The simplest type of swaging tool is a punch-like tool that can be forced into the end of the tubing to expand it to the size of the tool.
    c. Swaging tools (Figure 13-52) can be hammered into the end of tubing.
    d. Do not use swaging tools as punches or drifts; the smooth surface of the tool must be maintained.

SLIDE 13-30

Summary

The most important rule to remember about tools is to always use the right tool for the job. That means make sure you aren't forcing a tool to do something it wasn't created to do. Walk the extra 10 steps back to the truck to get a hammer instead of pounding on something with your wrench. Don't try to force a wrench to loosen a nut if the wrench is the wrong size. Don't try to cut certain materials with saws that are not meant to cut that type of material. Your job will go more smoothly, and your finished product will be of higher quality if you will make sure that your tools are in good working order, if you follow all safety rules, and if you match the job to the tool and the tool to the job.

Chapter 14 Power Tools—Selection, Use, and Care

Instructor's Note

In this column, instructors will find a variety of lecture tips to aid student comprehension and notes on when to use PowerPoint slides to teach concepts.

Appendix B lists corresponding numbered PowerPoint Presentation titles

SLIDE 14-1 OBJECTIVES

OBJECTIVES

- Students will have a working knowledge of common power tools.
- Students will be able to explain the operations commonly done with each of the power tools discussed.
- Students will be able to perform simple operations with power tools.

Glossary of Terms

Ampere—unit of measure for electrical current. Also abbreviated amp. Many power tool motors are sized according to the amperage their motors draw.

Ball bearing—style of bearing in which moving parts roll on steel balls.

Chuck—the part of a drill that holds the drill bit.

Chuck key—a special tool used to tighten a drill chuck.

Combination blade—saw blade that can be used for ripping and crosscutting.

Double-insulated—a style of electric tool construction that shields the user from the electrical parts of the tool.

Plunge cut—a cut made by plunging the saw in the middle of the work piece instead of cutting in from an outside edge.

Pneumatic—powered by compressed air.

Polarized plug—an electrical plug with one prong wider than the other so that it can only be plugged into the receptacle one way.

Reciprocate—move back and forth in a straight line. Some saw blades reciprocate.

Reversing switch—most electric drills have a reversing switch that allows the user to reverse the direction of the drill.

Roller bearing—style of bearing in which moving parts roll on small steel rollers.

Shoe—as used on power tools, this refers to the part of the tool that rests on the work piece.

Skil saw—the first portable circular saw was made by Skil, so some people call all portable circular saws Skil saws.

Sleeve bearing—style of bearing in which moving parts ride on a smooth metal sleeve or tube.

SLIDE 14-2

A. **Power Tool Construction**
   1. Motors
      a. The size of the power tool motor is specified by the amount of electrical current it uses.
      b. Electric current is measured in amperes (amps).
         - A typical electric drill motor (Figure 14-1) might be 14 amps.
         - Higher amperage equates to more power but also more weight and higher cost.

2. Cordless Tools
    a. Rechargeable batteries supply the power for cordless tools.
    b. Having two batteries (Figure 14-2) allows one to be in use while the other is recharging.
    c. Cordless tools offer portability and safety; little or no shock danger.
3. Insulation and Grounding

SLIDE 14-3

    a. Double-insulation helps protect users from shock.
    b. Double-insulated power tools have two chassis.
        - One encases all of the electrical wires
        - Another insulates the user from the inner one.
    c. Power tools need to be connected to an electrical ground (Figure 14-3).
4. Pneumatic Tools
    a. These tools use compressed air for power.
        - Require air compressor
        - Require air hose
    b. Examples of pneumatic tools
        - Nailers
        - Staplers
        - Grinders
        - Saws
        - Drills
        - Sanders
        - Wrenches
5. Bearings
    a. Bearings used on rotating shafts greatly reduce friction buildup.
    b. Three types used in power tools (Figure 14-4)
        - Sleeve bearings have a smooth surface on the inside of a short sleeve; shorter life expectancy than other bearings.
        - Roller bearings have free-rolling, straight rollers that allow the shaft to roll in the support of the bearing.
        - Ball bearings are usually found where the bearing must resist end-to-end as well as side-to-side movement.

SLIDE 14-4

Teaching Tip: First and foremost make sure students understand that with power tools the focus on safety doubles. Now, they need to be aware of where their hands and feet are, that they wear no loose clothing, they DO use PPE, and are always aware of the potential for electrical shock.

B. Power Tool Safety

1. Safety Rules for all Power Tools
    a. Use power tools only after you have received instruction in their use.
    b. Tools that are not double-insulated must be connected to an electrical ground.
    c. Wear appropriate protective clothing when operating power tools, such as safety glasses and hearing protection.
    d. Keep all guards and protective devices in place.
    e. Do not use defective tools.
    f. Unplug corded tools and pneumatic tools when changing blades, bits, and attachments.
    g. Disconnect electric and pneumatic tools when they are left unattended.
    h. Pneumatic tools should be connected with a quick-disconnect coupler.
    i. Make sure that the cutting edge of the tool will have a clear path.

j. Plug corded tools only into GFCI-protected outlets.
k. Do not force the tool.
l. Do not operate electric power tools near flammable liquids or in a dangerous atmosphere.

C. Portable Circular Saw

SLIDE 14-5

1. The first portable circular saw was invented by a company named Skil (Figure 14-5).
2. Portable saw parts
   a. Motor
   b. Handle with a trigger switch
   c. Shoe that tilts up to 45° and moves up and down to vary the depth of the cut
   d. Blade
   e. Blade guard
   f. Various other features depending on model (Figure 14-6)
3. Portable saws are sized according to the diameter of their blade.
4. Different blades are available for different operations (Figure 14-7).
5. Inspection and defects

Teaching Tip: Demonstrating the various cuts and safe ways to handle portable saws will help students have the opportunity to see this done the right way.

   a. Before plugging the saw in check the cord, the plug, and the location where the cord attaches to the saw.
   b. Make sure that the blade is attached properly.
   c. Make sure the guard retracts completely and that nothing is loose.
6. Crosscutting (cutting off a piece of board)
   a. Make sure a crosscutting blade is installed and the teeth are pointing in the right direction.
   b. Adjust the shoe so that the blade penetrates about 1/8 inch to 1/4 inch through the stock.
   c. Check the blade guard to see that it works freely.
   d. Use sawhorses to support the piece to be cut so that none of the portion to be cut off is unsupported (Figure 14-8).
   e. Use a square and a sharp pencil to mark a line on the stock where it is to be cut.
   f. Plug the saw in.
   g Grasp the saw by the two handles, and turn it on.
   h. Rest the forward end of the shoe on the stock, and push it slowly, but firmly into the stock.
   i. After the cut is complete, release the trigger switch but hold the saw until the blade stops.
   j. Unplug the saw.
7. Ripping
   a. In ripping, the blade must be either a combination or a rip blade; most often the combination blade is used.
   b. Support the board being ripped in such a way that the blade will not hit the sawhorse or other support (Figure 14-9).
8. Plunge Cuts
   a. Carefully lay out the cut to be made; paying particular attention to anything that might be on the other side of the material.

b. Adjust the saw for the proper depth of cut (1/8 inch to 1/4 inch through the material).
c. Hold the guard open and tilt the saw up on the front of the shoe.
d. Move the saw so the blade lines up with where you want the cut to be made.
e. Start the saw; slowly lower the blade into the work (Figure 14-10).
f. Follow the line carefully until the entire shoe rests on the material.
g. Release the guard so that it covers the blade.
h. Push the saw through the cut until you reach the corner of the opening.
i. Release the trigger switch and wait for the saw to stop turning before lifting it out of the kerf.

SLIDE 14-6

D. Saber Saw

1. The saber saw (Figure 14-11) is used to make curved cuts and to make small openings.
2. Many saber saws have a variable-speed switch and can be used for both straight up and down strokes as well as orbital cuts.
3. The shoe can be tilted to make bevel cuts.
4. Many blades are available for cutting wood, metal, plastics, and other materials (Figure 14-12).
5. Inspect the saw.
   a. Check the power cord for broken insulation.
   b. Check the plug.
   c. Check to make sure the cord is firmly attached to the saw.
   d. Check for loose parts.
6. Using the Saber Saw
   a. Select the best blade for the type of cutting you will do.
   b. Mark the cut with a pencil.
   c. Wear safety glasses.
   d. Hold the shoe of the saw firmly on the work. With the blade clear, turn on the saw. If it is a variable-speed saw, set the switch for the best speed for your work.
   e. Slowly push the saw into the work; keep the kerf on the waste side of the pencil line.
   f. Maintain firm downward pressure on the saw to reduce vibrations and improve cutting speed.
   g. Turn the saw slowly as it advances to follow curves.
   h. Do not twist the saw without advancing. This will break the blade and the result is a rough cut.
   i. Do not force the saw into the work. Advance it only as fast as it will cut easily.
7. Making Plunge Cuts
   a. Use a pencil to mark where the cut is to be made with a pencil
   b. Make sure that nothing behind the material will be in the path of the blade.
   c. Tilt the saw up on the front edge of the shoe with the blade in line with the pencil mark and clear of the work (Figure 14-13).
   d. Turn the switch on, holding the base steady.
   e. Gradually tilt the saw down until the blade penetrates the work and the base rests firmly on it.

f. Cut along the line.
   - Stay on the waste side.
   - Stay as close as possible to the line until you reach the corner.

g. Back up about an inch, turn the corner by cutting a small arc, and cut along the other side and into the next corner (Figure 14-14).

h. Turn the saw around, and cut in the opposite direction to remove the small pieces left in the corners.

SLIDE 14-7

E. Reciprocating Saw

1. This saw (Figure 14-15) is more powerful than a saber saw.
   a. Blade moves back and forth like a saber saw.
   b. Definitely a two-hand saw.
2. A switch allows the user to select between a straight-line motion or an orbital motion.
3. Variety of blades available to meet needs of job
   a. Rough cuts in wood
   b. Smoother cuts in wood
   c. Metal
   d. Plastics
   e. Demolition work (Figure 14-16)
4. Inspect all parts of the saw for damage, cracks, and loose fittings.
5. Using a reciprocating saw
   a. Insert the right blade for the job in the saw; tighten it according to the manufacturer's instructions.
   b. Make sure that nothing will be in the path of the blade except what you intend to cut. A reciprocating saw will make quick work of a pipe or electrical cable.
   c. Make sure the work piece is held securely. The rapid reciprocating motion of the saw can shake a loose work piece.
   d. Hold the saw comfortably in both hands, and pull the trigger switch to turn the saw on; move the selector switch to the desired position.
   e. Place the shoe against the edge of the work piece, and advance the saw into the cut slowly.

SLIDE 14-8

Teaching Tip: If possible, allow students to drill various practice items. If you can do so, let them drill through metal, wood and other materials.

F. Drills

1. There are heavy duty and light duty drills, and all drills have a handle, a trigger switch, a reversing switch, and a chuck (Figure 14-17).
2. Most light-duty drills are 1/4-inch or 3/8-inch drills.
3. Heavy-duty drills are usually 3/8-inch or 1/2-inch drills.
4. The part that does the cutting is called a drill bit.
5. Examples of most often-used bit types
   a. Twist drill
   b. Masonry
   c. Spade (Figure 14-18)
6. Screwdriver bits are also made especially for use in light-duty drills and power screwdrivers.
7. As with all power tools, a complete inspection of all parts is necessary prior to plugging in and using the drill.
8. Using a power drill

   a. The work piece must be held securely. Use clamps or temporarily nail it in place.
   b. Check behind the work piece to ensure that the bit has a clear path.
   c. Place the proper bit in the chuck, and hand-tighten the chuck.
   d. Makes sure the reversing switch is set in the forward position.
   e. Make a small indentation or dimple where the center of the hole should be.
   f. Plug the drill in; pull the trigger switch.
   g. Place the point of the drill in the indentation made in step 5; apply moderate pressure.
   h. Be prepared for the bit to catch on the edge of the material as it passes through the far side of the hole. Try to maintain even pressure in a straight line, and keep the drill speed constant at this point.
   i. If you are using a twist drill bit, keep the drill bit turning as you withdraw it from the hole.

9. Using a Drill to Drive Screws
   a. Drills can be used as screwdrivers or nut drivers.
   b. A corded drill will have a screwdriver attachment that has a clutch.
   c. Most cordless drills have a clutch mechanism that the user can set to allow the drill to slip when it reaches a certain amount of torque (Figure 14-21).

10. Hammer Drills
   a. Hammer drills (Figure 14-22) are similar to other drills, except that they deliver as many as 50,000 hammer blows per minute to the drill point as the drill turns.
   b. Most popular sizes are 3/8 inch and 1/2 inch.

SLIDE 14-9

G. Torches

1. Torches are used to solder and braze copper pipe and fittings in plumbing and HVAC industries.
2. Three different types of torches, along with three different types of fuel gases are common in the industry.
3. One common type of torch (Figure 14-23) is a small self-contained unit that consists of a fuel cylinder with the torch head attached directly to the cylinder.
4. Cutting and welding torches that have a double hose connecting the torch to two cylinders of gas are often oxyacetylene torches.
5. Torch safety is very important.
   a. Know how to use the torch properly.
   b. Make sure you know how to turn the torch off quickly.
   c. Know the location of and know how to use the nearest fire extinguisher.
   d. Do not point the torch toward your face, any part of your body, or another person.
   e. Disconnect the cylinder when not in use.
   f. Be extra careful when using the torch outdoors on sunny days.
      - Bright light makes it hard to see the flame.
      - Wind can blow the flame back on the operator.
   g. Do not use a leaking, damaged, or malfunctioning torch.
   h. Never use a torch to strip lead-based paint indoors.

Teaching Tip: Emphasize safety when using torches of any kind.

   i. Heating a surface may cause heat to be conducted to another part of the surface that is not intended to be heated.
      - Can cause a fire hazard.
      - Excess heat can cause pressure to build up to dangerous levels in an enclosed container or pipe.
   j. Always wear protective gloves when doing hot work.
   k. Never use a torch near combustible materials. Be especially cautious around motor fuels.

6. Soldering a Pipe Fitting
   a. Cut pipe to length using a tubing cutter.
   b. Ream to remove any metal burrs.
   c. Dry fit the pipe to ensure it goes squarely into the fitting.
   d. Use steel wool or an emery cloth to thoroughly clean both the outside of the pipe and the inside of the fitting so it is as bright as a new penny.
   e. Do not touch the cleaned area with your fingers.
   f. Immediately apply flux (soldering flux is a paste-like cleaner for this purpose) to these areas.
   g. Insert the pipe into the fitting.
   h. Secure the pipe and fitting so it cannot move during the soldering process.
   i. Direct the flame at the fitting until the metal becomes hot enough to melt solder at contact.
   j. Apply solder to the joint, feeding the solder along the joint and slightly behind the torch tip. Capillary action will draw the solder into the joint.
   k. Continue heating until the solder flows smoothly around the joint and fills it completely. When the joint is full, the solder will form a tiny bead around the joint.
   l. After you have finished soldering, clear away all flux residue around the joint with a clean, damp cloth.
   m. Let the joint cool by itself. Caution: Dipping it in water to cool will cause the joint to crack.

SLIDE 14-10

Summary

The most important word that you should associate with power tools is *SAFETY*. Always double-check your power tools prior to doing a job. No amount of haste is worth losing a limb or receiving a life-threatening shock simply because you didn't check to make sure your tools were in good working condition. The power tools available to make your job go faster and easier include saws, drills, and torches. There are portable saws that allow you to crosscut, rip cut, and plunge cut. There are saber saws for cutting into walls and making curve cuts. A variety of heavy-duty and light-duty drills allows you to drive screws, drill holes, and hammer. Torches can be used to solder and braze fittings. Each of these tools, however, comes with specific parts that must be checked before being used, and none of these power tools should be used for jobs they are not created to do.

Chapter 15 Fasteners

Instructor's Note

In this column, instructors will find a variety of lecture tips to aid student comprehension and notes on when to use PowerPoint slides to teach concepts.

Appendix B lists corresponding numbered PowerPoint Presentation titles

SLIDE 15-1 OBJECTIVES

OBJECTIVES

- Students will be able to identify and describe common fasteners.
- Students will be able to select the correct fastener types and sizes for common construction jobs.
- Students will be able to install or use each of the common fastener types.

Glossary of Terms

Anchor—a device that can be driven into or set in concrete, masonry, or other material to provide a place to attach a bolt. There are several types of anchors.

Box nail—this nail has a thin shank like a finishing nail, but a flat head like a common nail. It is usually coated to prevent loosening.

Brad—a very short nail with a small head, used to fasten thin parts.

Cap screw—a small bolt that usually has a hexagonal head.

Carriage bolt—a large bolt used in wood. It has a smooth oval head and a section of square shank right below the head.

Clinching—bending the protruding part of a nail over to make a permanent fastening.

Common nail—the most common type of nail, it has a heavy shank and a flat head.

Deck screws—these are similar to drywall screws, except they are more corrosion resistant.

Drywall screw—this is a light-gauge screw with a Phillips or square-drive head, used for fastening drywall to framing. Drywall screws are not as strong nor as corrosion resistant as other screws.

Duplex nails—this is a common nail with two heads, so that one can be driven tight and the other is still exposed for removal.

Finishing nail—this has a thin shank and a small head that can be driven beneath the surface of the wood.

Lag screw—this is a large wood screw, with either a square or hexagonal.

Penny size (abbreviated *d*)—This is an old system of measurement that is still in use. It is from the number of pennies used to purchase 100 nails of that size. The higher the penny size, the longer the nail.

Pitch—this is a measurement of the number of threads in one inch of a screw or bolt.

Screw gauge—this is a number representing the thickness of a screw. The higher the gauge the thicker the screw.

Sheet metal screw—this is usually a self-tapping screw used for fastening sheet metal.

Stove bolt—a small bolt with a round or flat head and fitted for a screwdriver.

Tapcon—this is a concrete screw.

Toenail—a nail that is driven at an angle into the face of one piece to hold it to another piece. Toenailing usually requires at least one nail on each side of the piece being toenailed.

VIDEO #3
CHAPTER 15

SLIDE 15-2
SLIDE 15-3

Teaching Tip: Have a variety of fasteners on hand for students to handle during the discussion of each one. Give examples of ways you have used any of the fasteners.

A. Nails

1. Nails come in a wide variety of sizes, shapes, and uses (Figure 15-1).
2. Nail size is specified by penny size, referring to the long-ago practice of determining the size of a nail by how much a hundred nails of a certain type could be purchased for a certain amount of pennies.
3. Today, penny size refers to a nail's length (Figure 15-2).
4. Nails can be driven three ways (Figure 15-3)
   a. Face nailing: nail is driven straight through one member into another
   b. Toenailing: used when it is not possible to use face nailing and where extra resistance to being pulled apart is needed
   c. Clinching: driving nails through both members, then bending over the points to prevent withdrawal
5. Use nails about three times as long as the thickness of the piece to be nailed.
6. Select an appropriate hammer for the job to be done, such as a 20-ounce hammer to drive a 16d nail
7. Hold the hammer firmly, but without squeezing it, near the end of the handle (Figure 15-4).
8. If you bend the nail, pull it out and start over; it is very hard to drive a bent nail.
9. There are many types of nails.
   a. Common nails
      - Flat head
      - Smooth shank
      - Most often used of all nails
   b. Box nails
      - Similar to common nails but with a thinner shank and head
      - Less visible under paint
      - Usually coated, making them rust resistant
   c. Finishing nails
      - Thinner shank than common nails
      - Small head allows them to be driven below surface of wood
      - Ideal for installing trim, millwork where appearance is important
   d. Casing nails
      - Small, cone-shaped head
      - Can be driven below the surface of wood
      - Used for installing exterior door, window trim
   e. Duplex nails
      - Double heads
      - Used for temporary structures such as scaffolds
      - Lower head allows for a tight nailing attachment
      - Upper head allows for easy removal

f. Roofing nails
- Specified by length rather than penny size
- Have large heads
- Coated to prevent rusting

g. Masonry nails (Figure 15-5)
- May be cut nails or wire nails
- Made of hardened steel

SLIDE 15-4

B. Screws

1. Used when extra holding power is needed and in situations where parts may need to be disassembled.
2. Designated by type (determining the shape of the head and the pitch and/or depth of the threads), length, and gauge (gauge is a number indicating the diameter of the shank) (Figure 15-6 and Figure 15-7 and Figure 15-8).
3. Several varieties

a. Drywall screws
- Usually have a Phillips drive
- Sometimes called Twinfast screws
- Have a bugle head
- Range in length from 1 to 3 inches

b. Particleboard screws and Deck screws
- More corrosion resistant than Twinfast screws
- Particleboard screws are most common all-around wood screw for interior use
- Deck screws more common for exterior use

c. Sheet metal screws
- Used for fastening thin metal
- Some may be self-tapping
- Another type, called a tek, is self-drilling (Figure 14-9)
- Come in lengths from 1/4 inch up to several inches
- Available from 6 gauge to 12 gauge

d. Lag screws
- Large wood screws with square or hex heads
- Sized by diameter and length
- Two holes must be drilled for a lag screw
  1) One hole the size of the shank in the top piece
  2) Second hole smaller than the shank in the bottom piece allows threads to cut into the wood
- Washer should be used to prevent head from digging into wood

SLIDE 15-5

C. Bolts

1. Bolts have high holding strength and allow for easy disassembly.
2. Bolts are specified by type, diameter, and length.
3. Three types of bolts widely used in construction

a. Cap screws (Figure 15-10)
- Sometimes called machine bolts, but these aren't used on machines

Teaching Tip: Remember to discuss why a worker would select one fastener over another. You might also discuss how cost can become a factor in the fastener chosen.

- Available with hex heads, slotted heads, Phillips heads, internal square drive or Allen drive heads
- Variety of sizes (Figure 15-11)
- Available with either UNC or UNF threads

b. Stove bolts
- Small bolts with round or flat heads
- Threaded all the way to the head (Figure 15-12)
- Often used to join sheet metal parts
- Can have straight-slot or Phillips head
- Usually have UNC threads

c. Carriage bolts (Figure 15-13)
- Smooth, round head with a square section right below the head
- Used with wood
- Head of bolt grips the wood as the nut is tightened (use a washer below the tightening nut)

d. Washers
- Come in several types
  1) Flat washers
  2) Lock washers
- Used to protect parts being attached, keep tension on the nut

SLIDE 15-6

D. Anchors

1. Provide a means of attaching parts to concrete and masonry surfaces or to hollow walls
2. There are many types of anchors, but some common ones used in construction include

Teaching Tip: Have a variety of anchors on hand to show students.

a. Wedge anchors
- Heavy-duty anchors used to install machinery, hand rails, and anything that will put the anchor under a high load (Figure 15-15)
- Installation requires drilling a hole the same size as the anchor and driving the anchor into masonry or concrete.

b. Sleeve anchors (Figure 15-16)
- Inserted through the part to be anchored and into the hole in the concrete or masonry
- As anchor tightens, sleeve expands against inside of the hole

c. Split fast anchors (Figure 15-17)
- Medium-duty anchors
- Made of one-piece steel
- Two sheared, expanded halves at bottom

d. Double expansion anchors (Figure 15-18)
- Used with cap screws
- Placed in correctly drilled hole just below or flush with the surface
- Place part to be fastened and bolt it into place

e. Lag shields (Figure 15-19)
- Used with lag screws
- Inserted into a hole of the right diameter and at least 1/2 inch deeper than length of anchor

f. Tapcon
- Concrete screw (Figure 15-20)
- Widely used because it eliminates need for separate anchor and fastener
- Tapcons come packaged with the right size drill bit because size of the hole is important

g. Hammer-drive anchors (Figure 15-21)
- Have either zinc alloy body or nylon body and a steel expander pin
- Anchor inserted through the part to be fastened and into the masonry
- Allows for quick anchor that lacks the strength of anchors already mentioned.

h. Split ribbed plastic anchors (Figure 15-22)
- Also called conical screw anchors
- Commonly used for attaching light loads

SLIDE 15-7
SLIDE 15-8

E. Hollow-wall fasteners

1. Toggle bolts (Figure 15-23)
   a. Have spring-loaded wings that must be inserted past the wall surface
   b. Spring forces wings apart inside the wall
2. Plastic toggles (Figure 15-24)
   a. Four legs attached to a body with a hole through the body
   b. Fins on sides prevent it from turning in hole
3. Molly screws (Figure 15-25)
   a. Sometimes called hollow-wall expansion anchors
   b. Consist of expandable sleeve, machine screw, and fiber washer
4. E-Z Anchors (Figure 15-26)
   a. Have sharp point, deep threads
   b. Can be driven into gypsum wallboard with a Phillips screwdriver

SLIDE 15-9

Summary

There are as many different types of fasteners as there are jobs that require fasteners. Some fasteners are heavy duty, while others are for more lightweight duty. It is important to examine the job to be done and determine which nails, screws, bolts, washers, and anchors would be best for the various jobs to be completed. It is also just as important to know which fasteners are rust resistant, which ones are used when appearance is important, and which ones are used for items that are not to be disassembled versus those that are more temporary structures. Having a good working knowledge of which fastener works best in a given situation saves time and money and makes for a safer working environment, as well as a quality finished product.

Chapter 16 Rigging

Instructor's Note

In this column, instructors will find a variety of lecture tips to aid student comprehension and notes on when to use PowerPoint slides to teach concepts.

Appendix B lists corresponding numbered PowerPoint Presentation titles

OBJECTIVES

- Students will be able to safely rig a roof truss for hoisting by a crane.
- Students will be able to use standard hand signals for directing a crane operator.
- Students will recognize common defects and safety hazards in rigging equipment.

SLIDE 16-1 OBJECTIVES

Teaching Tip: Throughout this entire section, re-emphasize safety over and over again. Any stories you can share, from your own experience will help students understand how critical safety is when it comes to rigging.

Glossary of Terms

Lay—the direction in which the strands are wound around the core of a wire rope. Also, the distance along a rope in which a strand makes a complete turn around the core.

Rated capacity—the amount of weight the manufacturer has specified for the maximum load on a sling.

Tag line—light line attached to a load to control it by hand.

Thimble—steel insert in the eye of a wire rope sling to prevent kinking and wear.

Wire rope—a rope made from strands of wire wrapped around a core in a particular way. Wire rope is sometimes incorrectly called cable.

Wire rope clip—a clamping device used to hold two pieces of wire rope to form an eye.

SLIDE 16-2

A. Slings

1. Practically all slings are made of wire rope, synthetic webbing, or steel alloy chains and can be used in a variety of different ways (Figure 16-1).
2. All slings have a manufacturer's tag that shows:
   a. Manufacturer's name
   b. Rated capacity for hitches or angles at which sling should be used
   c. Type of material (type of material for synthetic slings, grade of steel and link size for chains, or size of wire rope)
3. Wear-pads protect the sling from sharp corners on the load.

SLIDE 16-3

B. Wire Rope

1. Wire rope is made up of wires laid around a center in a specified pattern in one or more layers to form a strand. These strands lay around a core to form a wire rope (Figure 16-2).
2. Wire rope made up of a few large wires will be more abrasion resistant and less fatigue resistant than a rope of the same size made up of strands with many smaller wires.

SLIDE 16-4

C. Wire Rope Cores

1. The primary function of the rope's core is to serve as the foundation for the strands.

2. The type of core has an effect on the rope's performance.
3. There are three common types of cores:
    a. Fiber core
        - Polypropylene (the kind of synthetic rope used to tow water skiers) is standard.
        - Fiber or other man-made fibers are available on special request.
    b. Independent wire rope core
        - Literally an independent wire rope with strands and a core called IWRC.
    c. Strand core
        - Strand made of wires
        - Usually used only in utility cables

SLIDE 16-5

D. Lay and Rope Design

1. Lay
    a. Refers to the wire and strand positions in the rope
    b. Also used as a length measurement in manufacturing and inspection (Figure 16-3)
    c. Refers to the direction the strands lay in the rope: right or left
    d. Refers to the relationship between the direction the strands lay in the rope and the direction wires lay in the strands.
        - In appearance, wires in regular lay run straight down the length of the rope.
        - In lang lay, they appear to angle across the rope.
        - In regular lay, wires are laid in the strand opposite the direction the strands lay in the rope
        - In lang lay, the wires are laid the same direction in the strand as the strands lay in the rope.
    e. The length along the rope that a strand makes one complete spiral around the rope core (Figure 16-4). This is a measurement frequently used in wire rope inspection. Standards and regulations require removal when a certain number of broken wires per rope lay are found.
2. Regular lay rope is more stable and more resistant to crushing than lang lay.
3. Lang lay is more fatigue resistant and abrasion resistant.

SLIDE 16-6

Teaching Tip: Discuss in detail the various things students will look for when inspecting wire rope, slings, and the hardware that goes with them.

E. Inspection of Wire Rope

1. Periodic inspections are critical as all wire ropes will eventually wear out and gradually lose their ability to do the work for which they are intended.
2. A record must be kept of each inspection (Figure 16-5).
3. These inspections should be carried out by someone trained in what to look for.
4. When a wire breaks under tensile load, a "cup and cone" look appears at the point of failure (Figure 16-6).
    a. A wire with a fatigue break (Figure 16-7) is recognized by the square end perpendicular to the wire.
    b. Fatigue failure of a wire rope subjected to heavy loads over sheaves appears as breaks in the valleys of the strands, caused by "strand nicking" (Figure 16-8).
    c. Strand nicking (Figure 16-9) is a result of adjacent strands rubbing against one another.

d. A "birdcage" (Figure 16-10) is caused by sudden release of tension and the resulting rebound of rope. The rope should be replaced immediately.

e. Figure 16-11 shows a wire rope with a high strand—a condition in which one or more strands are worn before adjoining strands.

f. A kinked wire rope (Figure 16-11) is caused by pulling down a loop in a slack line during handling, installation, or operation. This rope must be replaced.

SLIDE 16-7

F. Web Slings

1. Synthetic web slings (Figure 16-13) offer several advantages over wire rope.
2. They conform to the shape of the load, so they are less apt to damage the load and less apt to slip.
3. They do not corrode, so they do not stain the surface of the load.
4. They can be damaged by exposure to temperatures over 180° and can deteriorate if left exposed to sunlight for long periods of time.
5. Like wire rope, a trained individual should inspect these slings frequently.

SLIDE 16-8
SLIDE 16-9

G. Construction

1. Web slings are made of nylon or polyester.
2. Polyester slings are damaged by caustics but resist acids.
3. Nylon slings are damaged by acids but resist caustics.
4. They are configured similarly and inspecting them is the same (Figure 16-14).
5. Web slings with hardware fittings usually have a male and a female end; the male end is passed through the female end for a choker hitch.
6. All synthetic slings have warning markers embedded in the material (Figure 16-15).

SLIDE 16-10

H. Inspection of Synthetic Slings

1. OSHA regulations require you to inspect your sling before each lift (Figure 16-16).
2. If you see damage, take the sling out of service.
3. There are reasons for removing a sling from service according to the American National Standards Institute (ANSI B30.9):
   a. Acid or caustic burns
   b. Melting or charring of any part of the sling
   c. Holes, tears, cuts, or snags
   d. Broken or worn stitching in load-bearing splices
   e. Excessive abrasive wear
   f. Knots in any part of the sling
   g. Excessive pitting or corrosion, or cracked, distorted, or broken fittings
   h. Other visible damage that casts doubt as to the strength of the sling
4. Slings should not be used any time you see any of the following:
   a. Red warning markers
   b. The sling is distorted in any way
   c. The identification tag is not readable
   d. Any time a sling is loaded beyond its rated capacity for whatever reason
5. Use good judgment when you visually inspect your slings; look for wear to the sling body, the selvage edge of webbing, and the condition of the sling eyes.

SLIDE 16-11
SLIDE 16-12

I. Chain Slings

1. Chain slings (Figure 16-17) are used for heavy loads or in conditions where heat would damage other types of sling materials.
2. Chain slings are not often used on construction sites due to potential damage they can cause to the load.
3. Anyone rigging a chain sling should be able to identify damage to the sling (Figure 16-18).

SLIDE 16-13

J. Effect of Angle on Sling Capacity

1. Using slings at an angle can be deadly if the angle is not taken into consideration (Figures 16.36 and 16.37).
2. Follow these steps to calculate the capacity of the sling needed for a job:
   a. Determine the weight that the sling will be lifting (LW).
   b. Calculate the tension factor (TF) by dividing the sling height (H) by the sling length (L).
   c. Multiply the lifting weight (LW) by the tension factor (TF) to find the minimum sling rating needed for that lift.

SLIDE 16-14
SLIDE 16-15

K. Hardware Attachments

1. Hooks are permanently attached to slings and are used to attach the sling to the load.
2. There are several types of hooks with specific purposes (Figure 16-38):
   a. Eye hooks have an eye for attachment of the sling or hoist.
   b. A reversed-eye hook has an eye at a right angle to the plane of the hook.
   c. A grab hook has a narrow throat so that it will grab a link in the chain and will not slip.
   d. A foundry hook has a greater throat opening, so the chain links or wire rope will slide through the hook.
   e. Many hooks used for slings are latch hooks that have a latch that helps prevent the load from slipping out of the hook.
   f. A clevis hook has a clevis-style pin instead of an eye for attachment to the hoist or sling.
   g. A choker hook has a barrel-like opening through which a wire rope can slide when used in a choker hitch.

L. Other Sling Hardware

1. Shackles (Figure 16-39) are used to attach a sling to a load or to join slings together.
2. Lifting eyes (Figure 16-40) can be straight eyebolts or swivel eyes.
3. Straight eyebolts should only be used where the load will pull straight along the axis of the eyebolt.
4. Lifting eyes should be inspected for:
   a. Wear that has removed part of the material
   b. Bent shank or eye
   c. Stress cracks
   d. Rust or corrosion
   e. Damaged threads

SLIDE 16-16

M. Thimbles

1. A thimble is used inside the eye of a wire rope sling to decrease wear and to protect the wire rope from kinking (Figure 16-41).

2. The thimble is inspected for wear, rust, or other visible damage each time the sling is inspected.

N. Wire Rope Clips

1. These are used to fasten a loop in a wire rope (Figure 16-42).
2. Clips come in various sizes to match the rope with which they will be used.
3. Use torque wrenches to tighten the clips.

SLIDE 16-17

O. Crane Operations

Teaching Tip: Crane safety and rules of operation are critical on the job site. Make sure students understand that they must be aware of their surroundings at all times on the job site, and especially when there is a crane in operation.

1. Follow basic safety rules when operating a crane to protect everyone at the job site.
   a. One person on the ground should be designated to direct the operations.
   b. Anyone who is not directly involved in the operation should be cleared from the path of the load and cleared from the area of the crane.
   c. Only the designated person should give signals to the crane operator.
   d. Use tag lines to control any load that might swing or be blown by the wind (Figure 16-43).
   e. Never attempt to hold a load or position it with your hands.
2. Prepare ahead of time the area where the load is going.
3. Any necessary blocking should be placed prior to lifting the load.
4. The only person in the landing area should be the person controlling the tag line to position the load as it comes to rest.

SLIDES 16-18 TO 16-32

5. Hand signals approved by the American National Standards Institute (Figure 16-44) should go from the rigger to the operator, or a radio can be used.

SLIDE 16-33

Summary

The use of slings, ropes, straps, chains, and fittings to hoist heavy objects into place with a crane is called rigging. Rigging involves more than just lifting heavy things, though. This is another place where safety is a must. Heavy items moving through the air can cause serious injury if all safety procedures are not followed. Another key safety element is taking the time to carefully inspect the slings, ropes, and fittings for wear and damage. Knowing which type of sling to use in specific situations will make your job easier and safer. This also goes for selecting the right type of rope, bolts, and clips. A complete knowledge of what you are working with, and what the capacity of those items is, will be another important factor in successfully hoisting and moving heavy items on a job site.

Chapter 17 Views

Instructor's Note

In this column, instructors will find a variety of lecture tips to aid student comprehension and notes on when to use PowerPoint slides to teach concepts.

Appendix B lists corresponding numbered PowerPoint Presentation titles

OBJECTIVES

- Students will be able to recognize oblique, isometric, and orthographic drawings.
- Students will be able to draw simple isometric sketches.
- Students will be able to identify plan views, elevations, and sections.

SLIDE 17-1
OBJECTIVES

Glossary of Terms

Elevation—a view of an object as seen from the side and showing the height of the object.

Isometric drawing—a drawing with horizontal lines drawn at an angle of 30°.

Oblique drawing—a drawing with one surface shown on the plane of the paper and the adjoining surface at an angle.

Orthographic projection—a style of drawing in which separate sides of an object are shown as if projected against the inside of a glass box.

Plan—a view of an object as seen from directly above.

Section—a view of an object as though the object had been cut in half.

SLIDE 17-2
SLIDE 17-3

A. Isometric Drawings

1. Vertical lines are drawn vertically and horizontal lines are drawn at an angle of 30° from horizontal (Figure 17-1).
2. All lines are drawn in proportion to their actual length.
3. These drawings are often used to show construction details (Figure 17-2).
4. There are five steps to take in creating an isometric drawing, such as the one in Figure 17.3:
   a. Sketch a Y with the top lines about 30° from horizontal.
   b. Sketch the bottom edges parallel to the top edges.
   c. Mark off the width on the left top and bottom edges. This will be about twice the height.
   d. Mark off the length on the right top and bottom edges. The length will be about twice the width.
   e. Sketch the two remaining vertical lines and the back edges.
5. Other isometric shapes can be made by adding to or subtracting from this basic isometric brick (Figure 17-4).
6. Angled surfaces are sketched by locating their edges and then connecting them.

Teaching Tip: Guide students through the drawing process for each of the types of drawings they will be making and reading.

SLIDES 17-4 TO 17-6

VIDEO #4
CHAPTER 17

SLIDE 17-7
SLIDE 17-8

B. Oblique Drawings

1. The most irregular surface is drawn in proportion as though it were flat against the drawing surface.

2. Parallel lines are added to show the depth of the drawing (Figure 17-5).

SLIDE 17-9
SLIDE 17-10

C. **Orthographic Projection**

1. Used to show all information accurately and to keep all lines and angles in proportion.
2. The best way to describe this type of drawing is to imagine the object inside a glass box.
3. This glass box image allows for the corners and the lines representing the edges of the object to be projected onto the sides of the box (Figure 17-6).
4. If the box is unfolded, the images projected onto its sides will be on a single plane, as on a sheet of paper (Figure 17-7).
5. All surfaces that are parallel to the plane of projection are shown in proportion to their actual size and shape.
6. Surfaces that are not parallel to the plane of projection are not shown in proportion (Figure 17-8).

SLIDES 17-11 TO 17-14

D. **Plans, Elevations, and Sections**

1. A plan view of a construction drawing shows the layout of the object as viewed from above (Figure 17-9).
2. Elevations are drawings that show height and are also used to show cabinets and interior features (Figure 17-10).
3. A section view, usually referred to simply as a *section*, shows what would be exposed if a cut were made through the object (Figure 17-11).
4. A floor plan is a type of section view (Figure 17-12).

SLIDE 17-15

Summary

There are a variety of ways to show drawings of buildings and their parts. Isometric views show three sides in a single drawing. Orthographic drawings or projections show the sides of the object as though viewed straight on. Oblique drawings show one surface on the plane of the paper and the adjoining surface at an angle. You will see plans, elevations, and sections as well. It will be increasingly important for you to learn to read and even draw these views yourself in order to accurately communicate what the job calls for.

Chapter 18 Scales

Instructor's Note

In this column, instructors will find a variety of lecture tips to aid student comprehension and notes on when to use PowerPoint slides to teach concepts.

Appendix B lists corresponding numbered PowerPoint Presentation titles

OBJECTIVES

- Students will be able to identify the scale used on a construction drawing.
- Students will be able to read an architect's scale.

Glossary of Terms

Scale—The amount of reduction or enlargement of an object that is drawn other than actual size. Also, a device used to measure objects that are drawn proportionately smaller or larger.

VIDEO #4
CHAPTER 18

SLIDE 18-1
OBJECTIVES
SLIDES 18-2 TO 18-3

Teaching Tip: Scale drawings may be a little difficult to comprehend, at first. Go through this as slowly as is necessary to make sure all students can read drawings that are made to scale.

A. Scale Drawings

1. In order to draw an entire building on paper, a scale smaller than actual size is used.
2. Floor plans for a house are frequently drawn at a scale of 1/4 inch equals 1 foot (Figure 18-1).
3. The detail for the firewall in Figure 18-2 was drawn to scale.
4. The scale to which the drawing is made is noted on the drawing, usually alongside or beneath the title of the view.
5. Some graphic scales have graduations representing feet and inches drawn to the scale of the view (Figure 18-3).
6. The graduations on the scale indicator can be marked on the edge of a sheet of paper, then stepped off on the drawing (Figure 18-4).
7. The graduations may also be transferred to the drawing with dividers (Figure 18-5).

SLIDE 18-4

Teaching Tip: You should have an architect's scale for students to see and handle as you go through that section. Have them practice reading it. This might be a good time to do a little group work, reading the scale and working out some math problems. This real-life application of their math skills will be a good double check for you to see if you need to re-teach any of those earlier chapters.

B. Reading an Architect's Scale

1. All necessary dimensions should be shown on the drawing.
2. The instrument used to make drawings to scale is called an *architect's scale* (Figure 18-6).
3. The triangular scale includes 11 scales frequently used on drawings.
4. Two scales are combined on each face except for the full-size scale, which is fully divided into sixteenths.
5. The combined scales work together because one is twice as large as the other, and their zero points and extra divided units are on opposite ends of the scale.
6. The fraction, or number, near the zero at each end of the scale indicates the unit length in inches used on the drawing to represent one foot of the actual building.
7. The extra unit near the zero end of the scale is subdivided into twelfths of a foot (inches) as well as fractions of inches on the larger scales.

8. Reading an architect's scale is a multistep process.
   a. Turn the scale to the 1/4-inch scale.
   b. Counting the marks from the zero toward the 1/4 mark, there are 12 lines marked on the scale.
   c. Each one of these lines is one inch on the 1/4″ = 1′-0″ scale.
   d. The fraction 1/8 is on the opposite end of the same scale (Figure 18-7).
   e. This is the 1/8-inch scale and is read in the opposite direction.
   f. Notice that the divided unit is only half as large as the one on the 1/4-inch end of the scale.
   g. Counting the lines from zero toward the 1/8 mark, there are only six lines. This means that each line represents two inches at the 1/8-inch scale.
   h. Looking at the 1-1/2-inch scale (Figure 18-8), note that the divided unit is broken into twelfths of a foot (inches) and also fractional parts of an inch.
   i. Reading from the zero toward the number 1-1/2, notice the figures 3, 6, and 9. These figures represent the measurements of 3 inches, 6 inches, and 9 inches at the 1-1/2″ = 1′-0″ scale.
   j. From the zero to the first long mark that represents one inch (which is the same length as the mark shown at 3) and 4 lines. This means that each line on the scale is equal to 1/4 of an inch.
   k. Reading from the zero to the 3, read each line as follows: 1/4, 1/2, 3/4, 1, 1-1/4, 1-1/2, 1-3/4, 2, 2-1/4, 2-1/2, 2-3/4, and 3 inches.

SLIDE 18-5

Summary

The term *scale* is used both to indicate the tool used to measure drawings that are done to scale, as well as to refer to the amount of reduction or enlargement of a drawing to make it fit the page or show the necessary details. Almost all of the construction drawings you will encounter will be drawn to scale or smaller than actual size. It is important for you to understand the conventional scales used to make construction drawings and to be able to read the tools used to measure scaled drawings.

Chapter 19 Alphabet of Lines

Instructor's Note

In this column, instructors will find a variety of lecture tips to aid student comprehension and notes on when to use PowerPoint slides to teach concepts.

Appendix B lists corresponding numbered PowerPoint Presentation titles

SLIDE 19-1
OBJECTIVES

OBJECTIVES

Students will be able to distinguish between and understand the meanings of:

- Object lines
- Dashed lines (hidden and phantom)
- Extension lines and dimension lines
- Centerlines
- Leaders
- Cutting-plane lines

Glossary of Terms

Break line—a line used to indicate that an object continues beyond the limits of the drawing.

Centerline—a line used to indicate the center axis of an object.

Cutting-Plane Line—a line used to show from where a section view is taken and in what direction it is viewed.

Dimension line—a line used to indicate the size of an object or feature.

Extension line—a line used to indicate the extent of a dimension line.

Hidden line—a dashed line used to show edges that are hidden from normal view.

Leader—a line used to associate a label with an object. Leaders usually have arrow heads.

Object line—a heavy solid line used to show the outline and or shape of an object.

SLIDE 19-2

VIDEO #4
CHAPTER 19

SLIDE 19-3

A. Object Lines

1. Object lines are used to show the shape of an object (Figure 19-2).
2. These are solid lines, but it is necessary to keep in mind that other things are also drawn in with solid lines.

Teaching Tip: Have students practice drawing a variety of lines, as you call them out. Being able to draw them for themselves will increase their familiarity with the types, thickness, and use of the various lines.

B. Dashed Lines

1. One type of dashed line, the *hidden line*, is used to show the edges of objects that would not otherwise be visible in the view shown.
2. Hidden lines are drawn as a series of evenly sized short dashes (Figure 19-3). Only the most important things are shown with hidden lines.
3. Another type of dashed line is used to show important overhead construction (Figure 19-4). These dashed lines are called *phantom lines*, and they show objects that are not actually in the view.
4. Phantom lines are also used to show alternate positions of objects (Figure 19-5).

SLIDES 19-4 TO 19-6

5. To avoid confusion, the dashed lines may be made up of different weights and different length dashes, depending on the purpose (Figure 19-6).

SLIDE 19-7
SLIDE 19-8

C. Extension Lines and Dimension Lines

1. Extension lines are thin, solid lines that project from an object to show the extent or limits of a dimension (Figure 19-7).
2. Dimension lines are solid lines of the same weight as extension lines.
3. A dimension line is drawn from one extension line to the next, and the dimension is lettered above the dimension line.
4. On construction drawings, dimensions are expressed in feet and inches.
5. The ends of dimension lines are drawn in one of three ways (Figure 19-8).
6. Dimensions that can be added together to come up with one overall dimension are called *chain dimensions*.

SLIDES 19-9 TO 19-12

D. Centerlines

1. Centerlines are made up of long and short dashes.
2. Centerlines are used to show the centers of round or cylindrical objects.
3. Centerlines are also used to indicate that an object is symmetrical or the same on both sides of the center (Figure 19-9).
4. The center of a round object is shown by using two centerlines so that the short dashes cross in the center (Figure 19-10).
5. On construction drawings, the center of an arc is shown by crossing centerlines. The radius is dimensioned on a thin line from the center to the edge of the arc (Figure 19-11).

SLIDE 19-13

E. Leaders

1. Some dimensions are too small to be accurately shown in the drawing.
2. Leaders are the lines that show where these objects, made on a clear area of the drawing, belong.

SLIDE 19-14
SLIDE 19-15

F. Cutting-Plane Lines

1. A cutting-plane line is usually a heavy line with long dashes and pairs of short dashes, having identification at their ends.
2. Some section views may not be referenced by a cutting-plane line on any other view. These are *typical sections* that would be the same if drawn from an imaginary cut in any part of the building (Figure 19-14).

SLIDE 19-16

Summary

Drawings used in construction are used to communicate information through a language of lines, referred to as an alphabet of lines, since these lines serve as the letters that put the symbolic words (drawings) on paper. The weight of the thickness of lines is sometimes varied to show their relative importance. This difference in line weight sometimes helps distinguish the basic shape of an object from surface details.

Chapter 20 Use of Symbols

Instructor's Note

In this column, instructors will find a variety of lecture tips to aid student comprehension and notes on when to use PowerPoint slides to teach concepts.

Appendix B lists corresponding numbered PowerPoint Presentation titles

SLIDE 20-1
OBJECTIVES

OBJECTIVES

- Students will be able to identify and understand the meaning of door and window symbols, as well as material symbols.
- Students will be able to identify and understand the meaning of electrical and mechanical symbols, reference marks for coordinating drawings, and abbreviations.

Glossary of Terms

Abbreviations—shortened forms of words used on drawings.

Hash Mark—represents the joint between the pipe and the fitting.

Nominal Dimension—an approximate size that may not represent any of the actual dimensions of the unit. Nominal Dimensions are usually rounded off to whole inches or feet and inches and are used only as a convenient way to refer to the window or door size.

Reference Marks—also referred to as *callouts*, are marks on drawings that indicate where details or sections of important features have been drawn.

SLIDES 20-2 TO 20-10

Teaching Tip: Have students practice drawing a variety of symbols, or at least practice identifying the various symbols, on copies of drawings. Any hands-on practice only helps to plant it in their memories better.

A. Door and Window Symbols

1. Door and window symbols show the type of door or window used and the direction the door or window opens.
2. Three basic ways for household doors to open
   a. Swing
   b. Slide
   c. Fold (Figure 20-1)
3. The direction a swing-type door opens is shown by an arc representing the path of the door.
4. The seven basic types of windows are named according to how those windows open.
5. The sizes of windows and doors are usually shown on a special window or door schedules, but they might also be indicated by notes on the plans near their symbols.
6. Manufacturer's catalogs usually list several sets of dimensions for every window model (Figure 20-3).
7. The notations on plans and schedules usually indicate nominal dimensions.

SLIDE 20-11
SLIDE 20-12

B. Materials Symbols

1. The outline of an object drawing may be filled in with a material symbol to show what the object is made of (Figure 20-4).

2. Many materials are represented by one symbol in elevations and another symbol in sections. Examples: concrete block, brick.
3. When a large area is made up of one material, it is common to only draw the symbol in a part of the area (Figure 20-5).

SLIDES 20-13 TO 20-15

C. Electrical and Mechanical Symbols

1. Electrical and mechanical systems in a building include:
   a. Wiring
   b. Electrical devices
   c. Piping
   d. Pipe fittings
   e. Plumbing fixtures
   f. Registers
   g. Heating and air-conditioning ducts
2. Wiring is indicated by lines that show how devices are connected.
3. These lines are not shown in their actual position, they simply indicate which switches control which lights, for example (Figure 20-6).
4. Mechanical systems—plumbing and HVAC (heating, ventilating, and air conditioning)—are not usually shown in much detail on drawings for single-family homes.
5. Piping is shown by lines; different types of lines represent different kinds of piping.
6. Symbols for pipe fittings are the same basic shape as the fittings they represent.
7. A short line or *hash mark* represents the joint between the pipe and the fitting.
8. Plumbing fixtures are drawn pretty much as the actual fixture appears (Figure 20-7).

SLIDES 20-16 TO 20-20

D. Reference Marks

1. *Callouts*, called *reference marks*, on plans and elevations indicate where details or sections of important features have been drawn.
2. To understand these reference marks you have to understand the numbering system.
   a. The simplest numbering system for drawings consists of numbering the drawing sheets and naming each of the views.
   b. Sheet 1 might include a site plan and foundation plan.
   c. Sheet 2 might include floor plans.
   d. Sheet 3 might show elevations.
3. Architectural drawings are labeled A-1, A-2, A-3, etc.
4. Electrical drawings are labeled E-1, E-2, E-3, etc. (Figure 20-8).
5. Most of the drawings for a house are architectural
   a. Drawing set is fairly small
   b. Letters indicating the type of drawing are not usually included
   c. Views are numbered; second number shows on which sheet it appears (Example: 4/3, 4.3, or 4-3).
6. Numbering each view and the sheet on which it appears makes it easy to reference a section or detail to another drawing.
7. The identification of a section view is given with the cutting-plane line showing from where it is taken (Figures 20.9; 20.10; 20.11; 20.12).

Teaching Tip: If possible, bring in a series of drawings, depicting the information given in this section. Let students see how many pages are involved in a simple one-story, frame constructed home.

SLIDE 20-21

E. Abbreviations

1. Notes and labels of parts in construction drawings are usually abbreviated as much as possible to avoid crowding the drawing.
2. The abbreviations used on drawings are usually a shortened form of the word and are easily understood (Figure 20-13).

SLIDE 20-22

Summary

The use of standard symbols makes for clear communication, no matter who will be trying to interpret a set of construction drawings. Many features of construction cannot be drawn exactly as they appear on the building. Therefore, standard symbols are used to show various materials, plumbing fixtures and fittings, electrical devices, windows, doors, and other common objects. It is a good idea to memorize the most basic, most common symbols and abbreviations to help you interpret the plans and drawings you will encounter in your job.

Chapter 21 Plan Views

Instructor's Note

In this column, instructors will find a variety of lecture tips to aid student comprehension and notes on when to use PowerPoint slides to teach concepts.

Appendix B lists corresponding numbered PowerPoint Presentation titles

OBJECTIVES

- Students will be able to explain the general kinds of information shown on site plans.
- Students will be able to explain the general kinds of information shown on foundation plans.
- Students will be able to explain the general kinds of information shown on floor plans.

SLIDE 21-1 OBJECTIVES

Glossary of Terms

Floor plan—shows size and locations of rooms and most features in the rooms or on the specified level of the building.

Foundation plan—shows layout and size of foundation.

Site plan—also called plot plan. Shows size of building site and location of building, driveways, sidewalks, and utilities on the site.

SLIDE 21-2

A. Site Plans

1. A site plan gives information about the site on which the building is to be constructed.
2. The boundaries of the site (property lines) are shown.
3. The property line is usually a heavy line with one or two short dashes between longer line segments.
4. Civil engineers draw up surveys and usually work with decimal parts of feet, rather than feet and inches.
5. Site dimensions are usually stated in tenths or hundredths of feet (Figure 21-1).
6. A symbol or arrow of some type indicates what compass direction the site faces.
7. The site plan also indicates where the building is positioned on the site. Anyone reading the site plan will have this basic information without referring to the other drawings.
8. If the finished site is to include walks, drives, or patios, they will also be described by their overall dimensions.

Teaching Tip: Any site, foundation, or floor plans that you can share with students will be helpful in their understanding of these drawings. They should be getting more familiar with reading drawings by now, but this would be a good spot to check for understanding.

SLIDES 21-3 TO 21-6

B. Foundation Plans

1. A foundation plan shows the foundations, walls, and any other structural work to be done below the living spaces.
2. There are two types of foundations that are commonly used in homes and other small buildings.
   a. One type of foundation has a concrete base, called the footing, supporting foundation walls (Figure 21-2).

b. Another type of foundation is the slab-on-grade that consists of a concrete slab placed directly on the soil with little or no other support.

VIDEO #4
CHAPTER 21

3. Slabs on grades are usually thickened at their edges and wherever they must support a heavy load (Figure 20-3).
4. When the footing-and-wall type foundation is used, girders are used to provide intermediate support to the structure above (Figure 20-4).
5. The foundation plan includes all of the dimensions necessary to lay out the footings and foundation walls.
6. Notes on the plan indicate areas that are not to be excavated, concrete-slab floors, and other important information about the foundation (Figure 20-5).

SLIDE 21-7

C. Floor Plans

1. A section view taken at a height shows the placement of walls, windows, doors, cabinets, and other important features.
2. A separate floor plan is included for each floor of the building.
3. The floor plans give you the building layout; that is, how to get from one room to another.

SLIDE 21-8

D. Dimensions

1. Dimensions are given for the sizes and locations of all walls, partitions, doors, windows, and other important features.
2. On frame constructions, exterior walls are usually dimensioned to the outside face of the wall framing.
3. Windows and doors may be dimensioned about their centerlines (Figure 20-6).
4. Solid masonry construction is dimensioned entirely to the face of the masonry (Figure 20-7).
5. Doors and windows are indicated on floor plans by their symbols. If the building includes stairs, they are included in the floor plan. If the ceiling is framed with joists, their size, direction, and spacing are shown on the floor plan.
6. Architectural features such as exposed beams, arches in doorways, or unusual roof lines may be shown by phantom lines.

SLIDE 21-9

Summary

The combination of site plans, foundation plans, and floor plans give an overall view of the construction site. Some of the plan views show an entire building. The ability to read, understand, and work with the various plans is what makes you a vital asset to the construction company you work for. It is always a good idea to familiarize yourself with the way the building you are constructing works. Know the doorways and windows and where things are supposed to go.

Chapter 22 Elevations

Instructor's Note

In this column, instructors will find a variety of lecture tips to aid student comprehension and notes on when to use PowerPoint slides to teach concepts.

Appendix B lists corresponding numbered PowerPoint Presentation titles

OBJECTIVES

- Students will be able to orient building elevations to building plans.
- Students will be able to explain the kinds of information shown on elevations.

Glossary of Terms

Elevations—the exterior elevation drawings of the building; also used to refer to height or elevation above sea level or a specified point.

Grade Line—a heavy line showing the surface of the ground.

Headers—the framing across the top of a window opening.

Orienting—determining the relationship of one drawing to another.

Plate—the uppermost framing member in the wall.

VIDEO #4
CHAPTER 22

SLIDE 22-1
OBJECTIVES
SLIDE 22-2

Teaching Tip: Explain the process of orienting, using the various drawings, and then have students do some of this. Use the building you are teaching in as your first work product, and then take advantage of exercises in the book.

SLIDES 22-3 TO 22-6

A. Orienting Elevations

1. Orienting is determining the relationship of one drawing to another, as in being able to picture how the elevation in the front relates to the front of the floor plan.
2. Elevations are named according to compass directions (Figure 22-2).
3. When the elevations are named according to compass directions, they can be oriented to the floor plan, foundation plan, and site plan by the north arrow on those plans (Figure 22-3).
4. When north will not be known, due to plans being sold through catalogs or other reasons, the elevations are named according to their position as you face the building (Figure 22-4).
5. The front of a building can be checked by locating the main entrance.

B. Information on Building Elevations

1. Building elevations are normally very simple and show the finished appearance of the building better than other views.
2. The elevations that show the final appearance will do so with solid lines; underground portion of the foundation is shown as hidden lines (Figure 22-5).
3. The surface of the ground is shown by a heavy sold line called a grade line.
4. Grade lines usually include one or more notes to indicate the elevation above sea level or another reference point (Figure 22-6).
5. Most important dimensions are included on the building elevations in a string at the end of one or more elevations (Figure 22-7).
   a. Thickness of footing
   b. Height of foundation walls

c. Top of foundation to finished first floor
d. Finished floor to ceiling or top of plate
e. Finished floor to bottom of window headers
f. Roof overhang at eaves

SLIDE 22-7

Summary

Drawings that show the height of objects are called elevations. Being able to orient elevations helps the construction worker know where the north side of a building is, where the front of the building is, and where other parts of the building are in relation to other parts of the building. It is important for the construction worker to be able to interpret the various lines depicting elevations, plates, and headers. Important dimensions will be indicated on the building elevations, such as thickness of footing, height of walls, finished floor to ceiling, finished floor to window headers, and roof overhang eaves.

Chapter 23 Sections and Details

Instructor's Note

In this column, instructors will find a variety of lecture tips to aid student comprehension and notes on when to use PowerPoint slides to teach concepts.

Appendix B lists corresponding numbered PowerPoint Presentation titles

SLIDE 23-1
OBJECTIVES

VIDEO #4
CHAPTER 23

SLIDE 23-2
SLIDE 23-3

Teaching Tip: Use this section to double check understanding in areas of reading symbols, lines and reference marks. Have students measure things and find the scale on these drawings, as they become familiar with these types of views and details.

OBJECTIVES

- Students will be able to find and explain information shown on section views.
- Students will be able to find and explain information shown on large-scale details.
- Students will be able to orient sections and details to the other plans and elevations.

Glossary of Terms

Longitudinal section—section view showing what would be seen at an imaginary cut through the length of the building or part.

Transverse section—section view showing what would be seen at an imaginary cut across the building or part.

Typical section—a section view that shows typical construction. Variations away from this detail would not be typical and would require special detail drawings.

A. Sections
1. The typical section may be a section view of one wall or it may be a full section of the building.
2. Full sections are named by the direction in which the imaginary cut is made.
3. A transverse section (Figure 23-2), sometimes called a cross section, is taken from an imaginary cut across the width of the building.
4. A full section taken from a lengthwise cut through the building is called a longitudinal section (Figure 23-3).
5. A typical wall section includes several kinds of information.
   a. Footing size and material
   b. Foundation wall thickness, height, and material
   c. Insulation, waterproofing, and interior finish for foundation walls
   d. Fill and waterproofing under concrete floors
   e. Concrete floor thickness, material, and reinforcement
   f. Sizes of floor framing materials
   g. Sizes of wall framing materials
   h. Wall covering and insulation
   i. Cornice construction—materials and sizes
   j. Ceiling construction and insulation
6. Section views are used to show any special construction that cannot be shown on normal plans and elevations (Figure 23-4).

7. A section view is *in plan* when it shows the interior of the fireplace as viewed from above.

SLIDE 23-4
SLIDE 23-5

B. Other Large-Scale Details

1. A large scale may be all that is needed to show the necessary details.
2. The most common examples of details shown in this way are on cabinet installation drawings (Figure 23-6).
3. Many details are best shown by combining elevations and sections or by using isometric drawings (Figures 23-7 and 23-8).
4. Another method of showing detail is an exploded view (Figure 23-9).

SLIDE 23-6

C. Orienting Sections and Details

1. Details and sections that refer to only one place in the building are identified by a reference mark.
2. Sections are usually referenced by a cutting-plane line with arrows on the ends of the line to indicate from what direction the imaginary cut is viewed.
3. A reference mark near the arrow indicates where the detail drawing is shown.
4. Reference marks used for orienting details may vary from one set of drawings to another.
5. Usually, a system of sheet numbers and view numbers is used.

SLIDE 23-7

Summary

As with all plans at construction sites, being able to find and explain information on section views is very important. A large scale may be all that is necessary for showing some details. Reference marks can be used to orient details on large scales.

Appendix A Answers to Chapter Review Questions

Chapter 1 Organization of the Industry

1. Unskilled

    Have not completed and are not enrolled in an apprenticeship program. Have little skill in the building trades. Work as helpers to those who are skilled.

    Example

    plumber's helper

    Skilled Trade

    Someone who has a high level of skill in a specific building trade. Most skilled workers have either completed an apprenticeship or they have studied a trade in a college or technical school.

    Example

    Carpenter

    Technical

    Work with engineers, architects, and other licensed professions. Usually has some college education in the construction industry.

    Example

    Estimator

    Professional

    Four or more years of college in a licensed profession.

    Example

    Engineer

2. a. apprenticeship

    b. technical college

3. Community college
4. Designing roof trusses, designing foundations, planning site drainage for a large development
5. architect
6. The general contractor signs a contract with the owner agreeing to construct whatever is described in the contract documents. The general contractor may sign contracts with several subcontractors to do various parts of the job. The general contractor is responsible to the owner, even if a subcontractor did the work.
7. Architect's or Engineer (depending on what the drawing is)
8. Sole proprietorship

    Business is owned by an individual

    Limited liability partnership

    Business is owned by two or more people. Each owner is liable only for the share of his or her investment.

    Corporation

    Business is owned by stockholders. Each owner buys shares of the business. Profit or loss is shared according to what percentage of the business is owned. Liability goes to the corporation, not the stockholders.

9. Sole proprietorship
10. The stock holders are not liable if the corporation is sued.
11. d. All of the above
12. d. All of the above
13. Unions negotiate with employers to establish a pay rate for their members.

    Influence government to do such as things as pass worker safety laws

    Encourage more construction

    Support technology that is good for construction

14. International Association of Bridge, Structural, Ornamental and Reinforcing Iron Workers

    International Association of Heat and Frost Insulators and Asbestos Workers

    International Brotherhood of Boilermakers, Iron Ship Builders, Blacksmiths, Forgers and Helpers

    International Brotherhood of Electrical Workers

    International Brotherhood of Teamsters

    International Union of Bricklayers and Allied Craftworkers

    International Union of Elevator Constructors

    International Union of Operating Engineers

    International Union of Painters and Allied Trades

    Laborers' International Union of North America

Operative Plasterers' and Cement Masons' International Association of the United States and Canada

Sheet Metal Workers' International Association

United Association of Journeyman and Apprentices of the Plumbing and Pipefitting Industry of the United States and Canada

United Brotherhood of Carpenters and Joiners of America

United Union of Roofers, Waterproofers and Allied Workers

Utility Workers Union of America

Chapter 2 Working in the Industry

1. You do
2. attitude
3. Each student will have a different definition, but it should essentially say that ethics has to do with how one determines what is right to do.
4. Knowledge that you did the right thing, employers recognize those who have good ethics, more job opportunities
5. Unethical behavior is depressing
6.
    - *Listening:* Team members listen to each others' ideas. They build on teammates' ideas.
    - *Questioning:* Team members ask each other sincere questions.
    - *Respect:* Team members respect one another's opinions. They encourage and support the ideas of others.
    - *Helping:* Team members help each other.
    - *Sharing:* Team members offer ideas to each other and tell each other what they have learned.
    - *Participation:* Team members contribute ideas, discuss them, and play an active role together in projects.
7. Speaking, listening, writing, reading, and body language
8. Ask them to tell you what you said.
9. Try to paraphrase what they have said.
10. Consider your reader, use good penmanship, use good grammar, spell words correctly
11.
    - Building specifications
    - Instructions for a new tool or material
    - Memos from senior managers explaining new personnel policies
12. Look the other person in the eye, look cheerful, keep your hands out of your pockets.
13. Learning does not occur only in school. There are many opportunities to learn in formal classes and outside of classes after graduation from school and completion of an apprenticeship.
14. As our world becomes more complex, the work environment requires more and more knowledge.

Chapter 3 Job Safety

1. Work place conditions and poor work practices
2. Many jobs involve working with tools that can present electrical or cutting hazards; many jobs involve dangerous materials, many of which do not at first appear dangerous; and some jobs can put you in a dangerous position if done incorrectly.
3. Prevention of fires, information about harmful vapors, first aid in case of swallowing or splashing in the eyes, hazards resulting from contact with the skin
4. Blade guards, belt or chain-drive guards, double-insulation
5. It could cause a person to do something that is unsafe, such as handling live electrical conductors when the person thought he had been told they were not live.
6. Some prescription medicines cause drowsiness or could slow your reactions.
7. The student might name any unsafe condition; for example, a frayed electrical cord. Do not use it. Report it to the foreman. (Multiple answers are possible.)
8. To keep workers safe
9. Material Safety Data Sheet
10.
    - Always keep chemicals and solvents in their original containers with the manufacturer's label in place. It is very dangerous to use chemicals without knowing what they are.
    - Read labels. Before you use even the most ordinary supplies, read the manufacturer's label completely. If there is any danger, the manufacturer usually explains it, how to avoid it, and what to do if an accident occurs.
    - Read the MSDS for each hazardous or questionable product with which you come in contact. The MSDS will give you much more information than the product label.

- Use chemicals and other supplies only for their intended purpose. There is a solvent, adhesive, coating, or other chemical designed for just about anything you might want to do. Why try to force another substance to do the job?
- Follow all recommended safety precautions and use the recommended PPE for every product and operation.

11. Report them to a foreman, supervisor, or safety committee.
12. Hard hat, hearing protection, safety glasses, gloves, steel-toe shoes, particulate mask, safety harness (Others can be mentioned.)
13. 70 decibels (approximately the amount of noise from a hair dryer)
14. fuel, oxygen, heat
15. A: ordinary materials like paper, cardboard, and wood

    B: flammable liquids

    C: electrical

    D: combustible metals
16. Class B, because the water will spread the fuel. Class C, because of the risk of electric shock.
17. It is marked on the label
18. The trench can cave in, and the weight of the soil can prevent a person from breathing.
19. 2 feet
20. Do not work in a trench unless a person who has been trained in trench safety certifies that it is safe. Use a trench box or appropriate shoring whenever there is the slightest risk of the sides of the trench collapsing.
21. voltage
22. Dry wood
23. Electrical current leaves the source (generator), travels through the conductor to the person, through the person to the wet ground, then through the ground back to the source.
24. There are 13 rules stated in this chapter:
    - Know the location of the electrical panel or disconnect before beginning the work.
    - Complete a CPR certification course. Encourage your co-workers to do the same.
    - Be certain that all of your tools, test equipment, power cords, etc., are in good condition and working order.
    - Do not use electrical devices in wet or damp locations unless they are specifically designed for this environment.
    - Should you need to use a ladder, use a fiberglass model. Wood and aluminum ladders can conduct electricity should they come in contact with a live wire.
    - Do not work on an energized circuit.
    - Before performing any type of work or act, think about the consequences of that work or act. Do not take chances or cut corners!
    - Always work carefully. Place yourself in the safest position to avoid slips, falls, or backing into energized parts.
    - Make certain you are satisfied that you are working in the safest environment possible. Do not continue if you feel uncomfortable due to actions of yourself or those of a co-worker.
    - Remove all jewelry (watches, rings, bracelets, etc.). Gold is an excellent conductor of electricity!
    - Discard all worn, defective, or otherwise unsafe items.
    - Deenergize, lock-out, and tag-out all electrical circuits upon which work is to be performed.
    - Verify that the circuit is deenergized before beginning your work.

Chapter 4 Safety with Ladders, Scaffolds, and Machines

1. Anyone who has been trained by a "competent person"
2. According to OSHA, "competent person" means someone who is capable of identifying existing and predictable hazards in the surroundings or working conditions that are unsanitary, hazardous, or dangerous to employees and who has the authorization to take corrective measures to eliminate them.
3. a. The scaffold should be built on a firm base, with appropriate base plates and sills.

   b. Scaffold poles, frames, uprights, etc. must be plumb and braced to prevent swaying and displacement.

   c. Frames and panels must be connected by cross, horizontal, or diagonal braces, alone or in combination, which secure vertical members together laterally.

   d. All brace connections must be secured to prevent dislodging.

e. Employees must be able to safely access any level of a scaffold that is 2 feet above or below an access point.

f. OSHA standards specifically forbid climbing cross-braces as a means of access.

g. Portable, hook-on, and attachable ladders must be positioned so as not to tip the scaffold.

h. Fall protection must be provided on any scaffold 10 feet or more above a lower level. Guardrail systems must be installed along all open sides and ends of platforms.

4. fiberglass
5. Never let your belt buckle pass beyond either ladder rail.
6. Never
7. Haul materials up on a line rather than carry them up an extension ladder.
8. The second step from the top
9. 5 feet
10. The safety rules given in this chapter are:
    - Do not operate engine-driven equipment until you have been instructed in its proper, safe use.
    - Keep motor fuels in approved containers that are made for that type of fuel.
    - Keep open flame and sources of sparks away from fuel.
    - Do not refuel a hot engine.
    - Do not use gasoline or diesel fuel as a solvent or cleaning agent.
    - Have a fire extinguisher for class B fires on hand; know how to use it.
    - Never operate an internal combustion engine indoors.
    - Keep engine-driven equipment in good condition. Do not remove guards, shields, or other devices intended to protect the user from moving parts.
    - Keep engines dry.
    - Do not operate an internal combustion engine without a working muffler, and if you must work near the engine, wear hearing protection.
    - Remove the spark plug wire(s) before doing any repairs to the engine or attached machinery.
11. All of the above are true statements.

Chapter 5 Whole Numbers

1. 21
2. 4,781
3. 1,151 pounds
4. 305 feet
5. 31 cubic yards
6. 56 feet
7. 33 feet
8. 37 hours
9. 77 feet
10. 190 amperes
11. 25 cubic yards
12. 736 feet
13. 78,000 BTUs
14. 17 gallons
15. 140 feet
16. 24
17. 1,954
18. 39 feet
19. 12 pounds
20. 15 cubic yards
21. 14 inches
22. 2,529 feet
23. 75 cubic yards
24. 2 feet
25. 5 feet
26. 5 feet
27. 325 feet
28. 15 feet
29. 9 feet
30. 964 pounds
31. 45
32. 66
33. 9,821 feet
34. 5,278 gallons
35. 216 degrees
36. 168 feet
37. $444
38. 81
39. 48 pounds
40. 1,250 feet
41. 675 watts
42. 32 feet

43. 18
44. 63 yards
45. $126
46. 3
47. 24
48. 12 feet
49. 358 pounds
50. 120 volts
51. 6
52. 3
53. 11 hours
54. 6
55. 6
56. 22 pounds
57. 50 cubic feet
58. 8
59. $18 per hour
60. 32 inches
61. 149 hours
62. 30 bundles
63. 18
64. 360 ounces or 22 pounds, 8 ounces
65. 4
66. $1.289.70
67. 6
68. 1 piece 80 inches long
69. 6 hours

Chapter 6 Decimals

1. 21.93
2. 541.375
3. 42.27
4. 610.788
5. 8.6
6. $60.94
7. 11.625 inches
8. 28.125 inches
9. 16.5 amps
10. $98.11
11. 743.4 ohms
12. 16.4 squares
13. 40.6 pounds
14. 243.3 square feet
15. 4.75 gallons
16. 12.5 pounds
17. $420.45
18. $61,379.83
19. $28,620.17
20. $634.55
21. $860.65
22. 0.61 amperes
23. 1.187 amperes
24. 0.0098 lbs. per cubic foot
25. 0.0186 lbs. per square inch
26. 146.5
27. 2,253.15
28. 0.354
29. 0.03
30. 3.370
31. 5.69 gallons
32. 10.32 hours
33. 26 ohms
34. $58.08
35. $978.75
36. $18,066.23
37. 0.03992 inch
38. 12 gauge
39. 0.45 inch
40. 0.358 inch
41. 240
42. 64 feet
43. 0.81 gallons
44. 41.112 inches
45. 0.237 yards
46. 5.2 inches
47. 15.44 or 16
48. 0.04 ohms
49. $26.81
50. 1.57 gallons
51. 0.097
52. $3.22
53. 84.05 lbs. per cubic foot
54. 8.42 hours
55. $121.80

Chapter 7
Fractions

1. 1
2. 5/8
3. 7/8 inch
4. 127/160 gallon
5. 13/24 hour
6. 4 5/8
7. 16 1/24 gallons
8. 1 29/32 inches
9. 2 13/24 pounds
10. 4 watts
11. 1/2
12. 13/16
13. 1 1/6 foot
14. 11 3/8 tons
15. 81 1/32 inches
16. 2 41/48 ohms
17. 4 77/240 ohms
18. 4 7/8 inches
19. 2 7/8 gallons
20. 111 1/12 hours
21. 282 19/24 hours
22. 51 23/24 hours
23. 110 11/12 feet
24. 13/16 inch
25. 2 1/16 inches
26. 21/80
27. 7/12
28. 2 1/5
29. 4 43/56
30. 121 11/12
31. 1 1/8
32. 6 2/5
33. 4 13/32
34. 27/172
35. 111/184
36. 14
37. 412 1/4 inches
38. 125 1/8 inches
39. 23 1/8 inches
40. $232.80
41. 4 amps
42. 572 1/4 inches
43. 23/36 inch
44. 13/20 inch
45. 8 inches
46. $711.56
47. $85.31
48. 3 1/2 hours
49. $613.55
50. 62.4 or 62 3/8 feet
51. 0.375
52. 0.75
53. 0.25
54. 0.556
55. 0.844
56. 23/32
57. 5/8
58. 11/16
59. 17/32
60. 35/64
61. 142.10 feet
62. 1.256 inches
63. no, it is within tolerance
64. 33.35 inches
65. 1 7/16 inches

Chapter 8
Linear Measure

1. 1/2 inch
2. 1 1/8 inch
3. 1 3/4 inch
4. 2 3/8 inches
5. 2 13/16 inches
6. 16 inches
7. 187 inches
8. 37 inches
9. 42 inches
10. 261 inches
11. 1 foot, 3 inches
12. 6 feet, 2 inches
13. 5 feet, 11 inches
14. 10 feet, 5 inches
15. 1 foot, 11 inches
16. 6 feet, 9 inches
17. 20 feet, 3 inches

18. 11 feet, 1 inch
19. 11 inches
20. 10 feet, 11 inches
21. 5 feet, 10 inches
22. 7 feet, 9 inches
23. 7 feet, 6 inches
24. 5 pieces. 6 feet, 9 inches must be cut off the last piece.
25. 22 feet, 7 inches
26. 18 feet, 10 inches
27. 11 feet, 8 inches
28. 28 feet, 0 inches
29. 6 pieces. 1 foot, 9 1/2 inches must be cut off the last piece.
30. 42
31. 20 millimeters
32. 100 meters
33. 3 meters
34. 1.1 kilometers
35. 22 centimeters
36. 136 feet
37. 56 feet
38. 32 inches
39. 25.1 inches
40. 440 feet

Chapter 9 Percent and Percentage

1. 580
2. 68.75
3. 70%
4. 3,060 square feet
5. 1,350 square feet
6. 87.5%
7. $141,250
8. 119.4 pounds
9. 78 square feet
10. 60.8% or 61%
11. $792.00 profit, $7,392.00 selling price
12. $150,000
13. 1.5%
14. $4,607.12
15. $257,005.03
16. $108 simple interest, $408 amount
17. $35.20 simple interest, $675.20 amount
18. $85.60 simple interest, $4,365.60 amount
19. $1,700.00 simple interest, $21,700 amount
20. $153.90 simple interest, $1,233.90 amount
21. $9,800.00
22. $3,982.50
23. $125,550.00
24. 5%
25. $8,600.00
26.

Payment	Interest	Payment on Principal	Balance
$920.00	$640.00	$280.00	$7,720.00
$920.00	$617.60	$302.40	$7,417.60
$920.00	$593.41	$326.59	$7,091.01
$920.00	$567.28	$352.72	$6,738.29
$920.00	$539.06	$380.94	$6,357.35

27. $1,642.65 principal, $2,957.35 interest

Chapter 10 Area and Volume

1. 121 square inches
2. 126 square feet
3. 36 square feet
4. 125 square feet
5. 49.875 square feet
6. 236 square feet
7. 207 square feet
8. 149 1/2 square feet
9. 1,038 square feet
10. 20.25 square inches
11. three
12. 2 1/2 gallons
13. 1,475 watts
14. 405 square feet
15. 840 square inches
16. 56 square feet
17. 252 square inches
18. 12,840 square feet
19. 1,224 square feet
20. 423 3/4 square feet
21. 113.1 square inches

22. 44.18 square feet or 6,361.8 square inches
23. 153.9 square meters
24. 3,067.97 square feet
25. 127.2 square inches
26. 4.516 square inches
27. 278.12 square inches
28. 38.48 square feet
29. 141.4 square inches
30. eight
31. 64 cubic feet
32. 80 cubic inches
33. 1,157.6 cubic feet
34. 1,716 cubic inches
35. 174 cubic inches
36. 1,000 cubic feet
37. 6 board feet
38. 53 1/3 or 53.33 board feet
39. 26 truck loads
40. 15 cubic inches

Chapter 11 Right Angles

1. 19.2
2. 24.7
3. 52.0
4. 7.3 feet
5. 14.7 feet
6. 16.58 feet or 16 feet, 7 inches
7. 23.24 feet or 23 feet, 3 inches
8. 42.55 feet or approximately 42 feet, 6 1/2 inches
9. 12.2 miles
10. 11 feet, 8 3/4 inches
11. 10 feet
12. 8 feet
13. 6 feet
14. 5 feet
15. 16 feet

Chapter 12 Combined Operations

1. 35.9 hours
2. See Figure 11-2 IG

FIGURE 12-2 IG

ESTIMATE

KOOL REFRIGERATION COMPANY
123 MAIN STREET
HOMETOWN, CA 12345

DATE: ______________
SALES PERSON: ____________
NO. 2404

EQUIPMENT & PARTS		LABOR		
DESCRIPTION/ PART NO.	PRICE	DESCRIPTION	HOURS	RATE
3-ton air conditioner	$1,469.00	Labor	15	$22.00
tubing, fittings, and hardware	$62.50			

EQUIP. & PARTS TOTAL	$1,531.50
TAX	61.26
LABOR TOTAL	330.00
TOTAL ESTIMATE	$1,922.76

3. Approximately $54.00. The answer will vary depending on how intermediate steps are rounded off.
4. 40 cubic yards. The answer will vary slightly, depending on how intermediate steps are rounded off.
5. 1,087 gallons
6. 11,084 pounds
7. 16 feet, 11 5/8 inches

Chapter 13 Hand Tools—Selection, Use, and Care

1. i
2. m
3. k
4. j
5. n
6. f
7. a
8. h
9. p
10. e
11. o
12. b

13. g
14. d
15. l
16. c
17. The rules given in this chapter are:
    - Never look directly into a laser beam.
    - Never view a laser beam with an optical instrument (builder's level).
    - When possible, set up the laser level so it is above or below eye level.
    - Turn the laser off when not in use.
    - Do not point a laser beam at another person.
18. The teeth on a crosscut saw form sharp points. The teeth on a rip saw are flat, forming little chisels.
19. Nails often fly out from under the hammer and could strike a worker in the eye.
20. a. 1/2 inch
    b. 1 1/8 inch
    c. 1 7/16 inch
    d. 15/16 inch
    e. 1 13/16 inch

Chapter 14 Power Tools—Selection, Use, and Care

1.

a. Use power tools only after you have received instruction in their use.

b. Tools that are not double-insulated must be connected to an electrical ground.

c. Wear appropriate protective clothing when operating power tools. This always includes safety glasses and includes hearing protection around tools that make loud noise. Do not wear loose clothes, jewelry, or hair that may be entangled in the tool.

d. Keep all guards and protective devices in place.

e. Do not use defective tools.

f. Unplug corded tools and pneumatic tools when changing blades, bits, and attachments.

g. Disconnect electric and pneumatic tools when they are left unattended.

h. Pneumatic tools should be connected with a quick-disconnect coupler.

i. Check to see that the cutting edge of the tool will have a clear path.

j. Only plug corded tools into GFCI-protected outlets. (See Chapter 4 for information about GFCI protection.)

k. Do not force the tool.

l. Do not operate electric power tools near flammable liquids or in a dangerous atmosphere. Even cordless tools can ignite flammable vapors.

2.

a. Combination circular saw blade:
Size: diameter
Use: general sawing in wood

b. Circular saw:
Size: diameter of blade
Use: all types of sawing in wood

c. Saber saw:
Size: amperage of motor
Use: sawing curves in a variety of materials

d. Reciprocating saw:
Size: amperage of motor
Use: Rough sawing in awkward places

e. Cordless drill
Size: voltage of battery and size of chuck
Use: all types of drilling

f. Twist drill bit
Size: diameter
Use: drilling small holes in wood, metal, and plastics

g. Hammer drill
Size: chuck size and amperage of motor
Use: drilling holes in concrete

Chapter 15 Fasteners

1. common nails
2. casing nails or finishing nails
3. roofing nails
4. finishing nails
5. hardened masonry nails
6. galvanizing
7. Unified National Coarse thread
8. lock washer
9. The threads on a sheet metal screw are shaped to grip the edge of thin sheet metal. The threads on a cap screw are intended to engage a nut.
10. stove bolts

11. d
12. f
13. c
14. b
15. e
16. a
17. 10d casing nail
18. 1/4-inch cap screw
19. The nails are brittle, so the heads break off easily when struck with a hammer. This can propel the head like a bullet.
20. It has a square portion on its shank, just beneath the head.

Chapter 16 Rigging

1. on the manufacturer's tag
2. on the manufacturer's tag
3. to protect the sling from sharp corners and edges on the load
4. Lay refers to the direction in which strands are twisted into the rope and to the distance a strand covers in making one complete turn around the rope.
5. sudden release of tension on the rope
6. broken strands, kinks, bird caging
7. lighter weight, more flexible, fit the shape of the load, not affected by water
8. manufacturer's label
9. The sling is worn out.
10. Any of the following are acceptable answers:
    - Acid or caustic burns
    - Melting or charring of any part of the sling
    - Holes, tears, cuts, or snags
    - Broken or worn stitching in load-bearing splices
    - Excessive abrasive wear
    - Knots in any part of the sling
    - Excessive pitting or corrosion, or cracked, distorted or broken fittings
    - Other visible damage that causes doubt as to the strength of the sling
    - Red warning markers are showing
    - The sling is distorted in any way
    - The identification tag is not readable
    - Any time a sling is loaded beyond its rated capacity for whatever reason
11. a competent person
12. Any of the following are acceptable answers:
    1. Gouged links
    2. Worn links
    3. Bent links
    4. Stretched links
    5. Damaged hooks and attachments
13. on the manufacturer's tag
14. inside an eye in a wire rope
15. 2,400 pounds
16. choker
17. bridle
18. lower the boom
19. raise the boom and lower the load
20. the rigger or director

Chapter 17 Views

1. a) oblique
   b) orthographic
2. a) section
   b) elevation
   c) plan
3. c and d
4. isometric
5. orthographic

Chapter 18 Scales

1. 20 feet, 0 inches
2. 22 feet, 6 inches
3. 5 feet, 0 inches
4. 7 feet, 6 inches
5. 11 feet, 11 inches
6. 2 feet, 3 inches
7. 2 feet, 6 inches
8. 5 feet, 11 inches
9. 1 foot, 3 inches
10. 2 feet, 9 1/2 inches

Chapter 19
Alphabet of Lines

1. outlines and edges of objects
2. a dashed line
3. a short thin line used to show what a dimension line refers to
4. Longer lines broken by short dashes. When used in a circle or arc, the short dashes in centerlines cross at the center of the circle.
5. It shows where the imaginary cut was made to produce the section view.

Chapter 20
Use of Symbols

1. a) It could be either metal (iron or steel) or brick
   b) insulation
   c) plywood
   d) cross section of rough lumber
   e) earth
   f) duplex electrical outlet
   g) gas pipe
   h) three-way switch
   i) water heater
   j) waterproof duplex outlet
2. a) gypsum board
   b) foundation
   c) finish floor
   d) on center
   e) reinforced
   f) exterior
   g) column
   h) diameter
   i) elevation
   j) concrete
3. drawing number 6 on sheet number 4
4. drawing number 5 on the third mechanical sheet

Chapter 21
Plan Views

1. from above
2. site plan or plot plan
3. north
4. foundation plan
5. floor plan
6. outside of the wall
7. floor plan
8. face of the wall

Chapter 21
Elevations

1. A drawing that shows the height of an object
2. grade line or surface of the ground
3. the portion of the building that is below the surface of the ground
4. The following six dimensions are listed in the chapter:
   - Thickness of footing
   - Height of foundation walls
   - Top of foundation to finished first floor
   - Finished floor to ceiling or top of plate
   - Finished floor to bottom of window headers
   - Roof overhang at eaves

Chapter 23
Sections and Details

1. a section view showing an imaginary cut across the width of a building
2. a section view showing an imaginary cut across the length of the building
3. The following are the items listed in the chapter:
   - Footing size and material
   - Foundation wall thickness, height, and material
   - Insulation, waterproofing, and interior finish for foundation walls
   - Fill and waterproofing under concrete floors
   - Concrete floor thickness, material, and reinforcement
   - Sizes of floor framing materials
   - Sizes of wall framing materials
   - Wall covering
   - Cornice construction—materials and sizes
   - Ceiling construction and insulation
4. construction details
5. by looking for reference marks and cutting plane lines

Appendix B PowerPoint Presentations List

Appendix B lists corresponding numbered PowerPoint Presentation chapters and titles.

Chapter 1

1-1	Objectives
1-2	Construction Occupation Categories
1-3	Unskilled or Semi-Skilled Labor
1-4	Skilled Trades
1-5	Technicians
1-6	Technical Careers
1-7	Design Management
1-8	Design and Construction Overview
1-9	Roof Trusses
1-10	Forms of Ownership
1-11	Corporation
1-12	Types of Codes
1-13	Zoning Ordinances
1-14	Structural Codes
1-15	Site Related Codes
1-16	Restrictive Codes
1-17	Model Codes
1-18	Safety Codes
1-19	Summary

Chapter 2

2-1	Objectives
2-2	Ethics
2-3	Working on a Team
2-4	Communication
2-5	Life-Long Learning
2-6	Summary

Chapter 3

3-1	Objectives
3-2	Accidents
3-3	OSHA
3-4	Safety Shoes
3-5	Hard Hats
3-6	Safety Glasses
3-7	Particulate Masks
3-8	Safety Harnesses
3-9	Ear Protection
3-10	Hazard Communication Standard
3-11	Fire
3-12	The Fire Triangle
3-13	The Fire Tetrahedron
3-14	Fire Extinguisher
3-15	Trench Safety
3-16	Material Handling
3-17	Electricity
3-18	Series Circuit (Illustration)
3-19	Extra Electrical Safety
3-20	Lock-out (Illustration)
3-21	Properly locked out circuit breaker (Illustration)
3-22	Three Separate Padlock (Illustration)
3-23	Front and back (Illustration)
3-24	Summary

Chapter 4

4-1	Objectives
4-2	Scaffolds
4-3	Pump Jack Scaffold
4-4	Ladder
4-5	Safe Climbing Habits
4-6	Safe Climbing Habits
4-7	Safe Climbing Habits
4-8	Safe Climbing Habits
4-9	Safe Climbing Habits
4-10	Safe Climbing Habits
4-11	Safe Climbing Habits
4-12	Internal Combustion Engine
4-13	Compressed Air
4-14	Summary

Chapter 5

5-1 Objectives
5-2 Addition
5-3 Subtraction
5-4 Multiplication
5-5 Division
5-6 Combined Operations
5-7 Summary

Chapter 6

6-1 Objectives
6-2 Decimals
6-3 Basic Principles
6-4 Decimal (Illustration)
6-5 Addition and Subtraction
6-6 Shingles (Illustration)
6-7 Multiplication of Decimals
6-8 Multiplication Example
6-9 Examples
6-10 Division of Decimal Fractions
6-11 Division Examples
6-12 Summary

Chapter 7

7-1 Fractions
7-2 Basic Principles of Fractions
7-3 Reading a Tape (Illustration)
7-4 Adding Fractions
7-5 Adding Mixed Numbers
7-6 Subtracting Fractions
7-7 Resistors in Series (Illustration)
7-8 Multiplying Fractions
7-9 Dividing Fractions
7-10 Decimal and Common Fraction Equivalents
7-11 Summary

Chapter 8

8-1 Objectives
8-2 U.S. Customary Scale
8-3 Fractions of an Inch (Illustration)
8-4 Nominal Dimensions
8-5 Inches (Illustration)
8-6 Combining Feet and Inches
8-7 Metric System
8-8 Perimeter Measure
8-9 Summary

Chapter 9

9-1 Objectives
9-2 Percent
9-3 Percent Formula
9-4 Percent
9-5 Percentage
9-6 Percentage
9-7 Percentage
9-8 Percentage
9-9 Interest
9-10 Summary

Chapter 10

10-1 Objectives
10-2 Area of Squares and Rectangles
10-3 Square Foot
10-4 U.S. Customary System
10-5 Metric System
10-6 Area of Triangles
10-7 Area of Circles
10-8 Volume of Rectangular Solids and Cubes
10-9 Summary

Chapter 11

11-1 Objectives
11-2 Basic Principles
11-3 Hypotenuse (Illustration)
11-4 Triangle (Illustration)
11-5 6-8-10 Method
11-6 Triangle
11-7 Summary

Chapter 12

12-1 Objectives
12-2 Basic Principles
12-3 Summary

Chapter 13

13-1 Objectives
13-2 Systems of Measurement
13-3 Tape Measure
13-4 Rafter Square
13-5 Stair Gauges
13-6 Speed Square
13-7 Combination Square
13-8 Chalk Line Reel

- 13-9 6–8–10–Method And Checking Diagonals
- 13-10 Triangle (Illustration)
- 13-11 Spirit Level
- 13-12 Builder's Level
- 13-13 Transit (Illustration)
- 13-14 Measuring Elevations
- 13-15 Laser Level
- 13-16 Laser Level (Illustration)
- 13-17 Plumb Bob
- 13-18 Hammers
- 13-19 Bars and Nail Pullers
- 13-20 Screwdrivers
- 13-21 Pliers
- 13-22 Wrenches
- 13-23 Hacksaw
- 13-24 Handsaws
- 13-25 Using a Handsaw
- 13-26 Utility Knife
- 13-27 Pipe and Tubing Cutters
- 13-28 Flaring Tools
- 13-29 Swaging Tools
- 13-30 Summary

Chapter 14

- 14-1 Objectives
- 14-2 Power Tool Construction
- 14-3 Three-Prong Plug (Illustration)
- 14-4 Power Tool Safety
- 14-5 Portable Circular Saw
- 14-6 Saber Saw
- 14-7 Reciprocating Saw
- 14-8 Drills
- 14-9 Torches
- 14-10 Summary

Chapter 15

- 15-1 Objectives
- 15-2 Nails
- 15-3 Penny Sizes of Nails (Illustration)
- 15-4 Screws
- 15-5 Bolts
- 15-6 Anchors
- 15-7 Hollow-Wall Fasteners
- 15-8 Toggle Bolt (Illustration)
- 15-9 Summary

Chapter 16

- 16-1 Objectives
- 16-2 Slings
- 16-3 Wire Rope
- 16-4 Wire Rope Cores
- 16-5 Lay and Rope Design
- 16-6 Inspection of Wire Rope
- 16-7 Web Slings
- 16-8 Construction
- 16-9 Standard Eye Sling (Illustration)
- 16-10 Inspection of Synthetic Slings
- 16-11 Chain Slings
- 16-12 Chain Sling (Illustration)
- 16-13 Effect of Angle on Sling Capacity
- 16-14 Hardware Attachments
- 16-15 Standard Eye Hook (Illustration)
- 16-16 Thimbles/Wire Rope Clips
- 16-17 Crane Operations
- 16-18 Stop (Illustration)
- 16-19 Dog Everything (Illustration)
- 16-20 Move Slowly (Illustration)
- 16-21 Hoist (Illustration)
- 16-22 Lower (Illustration)
- 16-23 Raise Boom (Illustration)
- 16-24 Lower Boom (Illustration)
- 16-25 Swing (Illustration)
- 16-26 Raise the Boom and Lower the Load (Illustration)
- 16-27 Lower the Boom and Raise the Load (Illustration)
- 16-28 Travel (Illustration)
- 16-29 Extend Boom (Illustration)
- 16-30 Retract Boom (Illustration)
- 16-31 Travel (Illustration)
- 16-32 Travel (Illustration)
- 16-33 Summary

Chapter 17

- 17-1 Objectives
- 17-2 Isometric Drawings
- 17-3 Isometric of a Building (Illustration)
- 17-4 Gable Roof Building (Illustration)
- 17-5 Ell-Shaped Building (Illustration)
- 17-6 Stepped Building (Illustration)
- 17-7 Oblique Drawings
- 17-8 Crown Molding (Illustration)

17-9 Orthographic Projection
17-10 Glass Box Unfolded (Illustration)
17-11 Plans, Elevations, and Sections
17-12 Building Elevation (Illustration)
17-13 Section of a Window Sash (Illustration)
17-14 Floor Plan (Illustration)
17-15 Summary

Chapter 18

18-1 Objectives
18-2 Scale Drawings
18-3 Graphic Scale
18-4 Reading and Architect's Scale
18-5 Summary

Chapter 19

19-1 Objectives
19-2 Alphabet of Lines (Illustration)
19-3 Object Lines
19-4 Dashed Lines
19-5 Dashed Lines (Illustration)
19-6 Phantom Lines (Illustration)
19-7 Extension Lines and Dimension Lines
19-8 Dimension-line Ends
19-9 Centerlines
19-10 Centerline (Illustration)
19-11 Center of round object (Illustration)
19-12 Showing arc radius (Illustration)
19-13 Leaders
19-14 Cutting-Plane Lines
19-15 Cutting-Plane (Illustration)
19-16 Summary

Chapter 20

20-1 Objectives
20-2 Door and Window Symbols
20-3 Door Types and Their Plan Symbols (Illustration)
20-4 Window Types (Illustration) Double Hung
20-5 Window Types (Illustration) Sliding
20-6 Window Types (Illustration) Awning
20-7 Window Types (Illustration) Casement
20-8 Window Types (Illustration) Hopper
20-9 Window Types (Illustration) Jalousie
20-10 Window Types (Illustration) Fixed
20-11 Material Symbols
20-12 Plan and Section Symbols (Illustration)
20-13 Electrical and Mechanical Symbols
20-14 Some Common Electrical Symbols (Illustration)
20-15 Some Common Plumbing Symbols (Illustration)
20-16 Reference Markers
20-17 Sheet A-4 (Illustration)
20-18 Sheet 5 (Illustration)
20-19 Drawing 4 (Illustration)
20-20 Roof Section (Illustration)
20-21 Abbreviations
20-22 Summary

Chapter 21

21-1 Objectives
21-2 Site Plans
21-3 Foundation Plans
21-4 Footing and Foundation Wall (Illustration)
21-5 Slab-on-Grade Foundation (Illustration)
21-6 Girder (Illustration)
21-7 Floor Plans
21-8 Dimensions
21-9 Summary

Chapter 22

22-1 Objectives
22-2 Orienting Elevations
22-3 Information on Building Elevations
22-4 Underground portion of the building (Illustration)
22-5 Site Elevation (Illustration)
22-6 Elevation Dimensions (Illustration)
22-7 Summary

Chapter 23

23-1 Objectives
23-2 Sections
23-3 Special Section of Ventilated Ridge (Illustration)
23-4 Other Large-Scale Details
23-5 Cabinet Elevations
23-6 Orienting Sections and Details
23-7 Summary

Appendix C Answers to Workbook

Chapter 1

Multiple Choice

1. C
2. A
3. C
4. D
5. D
6. B
7. C
8. C
9. B
10. D

Short Answer

11. unskilled labor, skilled labor, technician, professional
12. unskilled
13. The ones listed in the textbook are: carpenter, plumber, roofer, electrician, mason, HVAC technician, plasterer, tile setter, equipment operator, drywall installer, and painter
14. The ones listed in the textbook are: surveyor, estimator, drafter, expeditor, superintendent, inspector, planner.
15. A business that is owned and operated by one person.
16. A business that is owned by shareholders and run by directors the shareholders choose.
17. It is a code that can be adopted as is or with modifications by a government agency.
18. To ensure that all buildings in an area are built to be safe.
19. The International Building Code
20. administer a pension plan, run apprenticeship programs, negotiate with employers, influence governments to pass laws that are good for their members.

Chapter 2

Multiple Choice

1. A
2. B
3. D
4. C
5. D
6. B
7. A
8. C
9. D
10. A

Application

1. It's just going to be thrown away anyway. I am a foreman, so I should make all decisions about what's in my van. It is going to a good cause.
2. It is not Alicia's paint. Being in charge of a crew does not make you in charge of everything you come in contact with.
3. She should have asked Mr. Gonzales if it is okay to donate that paint to the shelter.
4. Read about the company and talk to people who have worked there. Make sure he has clean, neat clothes to wear to the interview. Prepare a neatly written summary of his experience and training for the job.
5. Look the interviewer in the eye. Keep his hands out of his pockets. Speak up. Use good grammar. Be cheerful.

Chapter 3

Multiple Choice

1. A
2. B
3. D
4. C
5. B
6. C
7. B
8. A
9. C
10. D
11. C
12. A
13. D

14. B

Short Answer

15. There are many possible answers. The example in the textbook is shoveling concrete, because of the danger of chemical burns.
16. six are given in the book: performing tasks without proper training, failing to read labels and instructions, ignoring safety devices, engaging in horseplay, poor communication, using alcohol and drugs.
17. six are given in the textbook: poor housekeeping, wet conditions, excessive heat or cold, poor lighting, poorly maintained or inoperative equipment, fire and explosion hazards.
18. five are given in the textbook: Read the OSHA poster which every employer is required to display; follow the employer's safety and health rules and wear or use all required gear and equipment; follow safe work practices for your job, as directed by your employer; report hazardous conditions to a supervisor or safety committee; report hazardous conditions to OSHA, if employers do not fix them.
19. Whenever it is possible to fall more than 6 feet.
20. 70 decibels
21. To describe hazardous materials risks and how to guard against them
22. B
23. It provides a path to ground without going through the user's body.
24. Lock it out and tag it out.

Matching

25. c
26. f
27. a
28. j
29. d
30. b
31. i
32. e
33. g
34. h

Chapter 4

Multiple Choice

1. C
2. B
3. A
4. B
5. C
6. A
7. D
8. D
9. C
10. D

Short Answer

11. there are 8 listed in the textbook: have a level, firm base; vertical members must be plumb and braced; frames and panels must be connected by cross, horizontal or diagonal bracing; bracing connections must be secured; safe access to any point 2 feet above or below an access point; no climbing on cross braces; portable and attachable ladders must be positioned so as not to tip the scaffold; approved guard rails or personal fall-arrest systems must be used above 10 feet.
12. second from top step
13. The bottom of the ladder should be 1 foot from the wall for every 4 feet of ladder length.
14. Read the label.
15. Until your belt buckle reaches the side rail.
16. The exhaust contains large amounts of carbon monoxide, which can accumulate in enclosed spaces. An open door is not adequate ventilation.
17. When it cools.
18. Never use gasoline as a solvent.
19. fire, carbon monoxide, moving parts, electric shock, hearing loss.
20. never

Chapter 5

Operations

1. 11
2. 11
3. 151
4. 930
5. 3
6. 398
7. 18
8. 1,476
9. 2
10. 24

Problems

11. 23 hours
12. $42

13. 85 two by fours
14. 900 square feet
15. 6 tons
16. 200 nails
17. 39 yards
18. $585
19. 144 inches
20. 88 inches
21. 48 pounds
22. 120 inches
23. 240 screws
24. 7 feet
25. 55 pieces
26. 20 lamps
27. 4 air conditioners
28. 333 cubic feet
29. 12 hangers
30. 96 bundles
31. 3 feet
32. 6 feet
33. 40 coils
34. 14 hours
35. 11 outlets

Chapter 6

Operations

1. 0.8
2. 57.72
3. 14.86
4. 10.0
5. 141.23
6. 88.775
7. 0.18
8. 4.5
9. 12.0
10. 0.5

Problems

11. $195.00
12. 151 ohms
13. 9.375 inches
14. 12.0 inches
15. 7.25 inches
16. $336.66
17. $301.21
18. $71.83
19. 8.0 inches
20. 1.7 amps
21. 0.17 psi
22. 1.95 feet
23. 7.65625 gallons
24. $209.88
25. 49 outlets
26. 4.3 hours
27. 0.106 inches
28. 9.6 yards
29. $276.59
30. 0.02402 inches
31. It is 0.00041 inch oversize for #12
32. 15 pieces
33. $55.10
34. $12.90
35. $965.77

Chapter 7

Least Common Denominators

1. 4/6 and 3/6
2. 15/20 and 4/20
3. 4/28 and 7/28
4. 20/30, 15/30, and 12/30
5. 35/80, 45/80, and 8/80

Operations

6. 4/4 or 1
7. 13/15
8. 1 17/20
9. 9 13/16
10. 9/16
11. 2 7/8
12. 12 3/4
13. 102 21/32
14. 2/9
15. 2 1/16
16. 2 85/224
17. 158 2/5
18. 3/16
19. 1 37/63
20. 3 131/168

Problems

21. 1 3/16 inches
22. 48 17/24 ohms

23. 6 15/16 inches
24. 20 13/24 feet
25. 14 3/4 inches
26. 2 1/12 gallons
27. 13 1/4 yards
28. 18 buckets
29. 255 1/12 hours
30. The plumbers worked 1 17/24 more hours.
31. The masons worked 6 1/4 hours more than the electricians.
32. $317 3/4 or $317.75
33. $537.69
34. 6 3/4 inches
35. 7 systems
36. 100 3/4 inches
37. 119 1/4 inches
38. 4 7/32 gallons
39. 32 1/2 receptacles
40. 5 10/13 receptacles
41. 0.888 inches
42. 99.925 inches
43. 3.825 inches
44. 183.8 inches
45. $312.64

Chapter 8

Reading a U.S. Customary Scale

1. 3/4 in.
2. 1 9/16 in.
3. 2 3/8 in.
4. 3 5/8 in.
5. 4 9/16 in.

Combining Feet and Inches

6. 29 in.
7. 57 in.
8. 75 in.
9. 97 in.
10. 94 in.
11. 2 ft. 1 in.
12. 3 ft. 5 in.
13. 8 ft. 9 in.
14. 5 ft. 4 in.
15. 1 ft. 7 in.
16. 3 ft. 9 in.
17. 8 ft. 0 in.
18. 11 ft. 1 in.
19. 2 ft. 9 in.
20. 4 ft. 7 in.

Problems

21. 7 ft. 10 in.
22. 4 ft. 8 in.
23. 7 pieces, 77 inches
24. 20 ft. 5 in.
25. 11 ft 10 in.
26. 80 in.
27. 10 ft. 8 in.
28. 8 pieces
29. 5 pieces, 1 ft.
30. 32 pieces
31. 16 ft 5 in.
32. 40 in.
33. 31.416 in.
34. 110 ft 6 in.
35. 73 ft 11 in.
36. 28 pieces
37. 113.1 inches
38. 146.5 inches
39. 141.8 inches
40. 6 pieces

Chapter 9

Simple Calculations

1. 75%
2. 115.2
3. 20
4. 34.8
5. 23.5%
6. 41
7. 43.8
8. 80%
9. 4,840
10. 765.6

Problems

11. $410.35
12. 40.9%
13. $119.04
14. 3146 square feet
15. $789.44
16. $71,900.00

17. $9,229.55
18. $101.86
19. 3.0%
20. 26.6%
21. 0.003 psi
22. $76,677.30
23. 49.1%
24. $495.00
25. 6 bundles
26. 2,083.3 pounds
27. 7.1%
28. 3 containers
29. $272.60
30. 35.4%
31. $6,600.00
32. $3,080.00
33. 3.5%
34. $41,500.00
35. $1,107.76

Chapter 10

Area

1. 16 square feet
2. 36 square inches
3. 576 square inches or 4 square feet
4. 2,048 square inches or 14 square feet
5. 3,136 square inches or 22 square feet
6. 328 square inches
7. 8 square feet
8. 1,008 square inches or 7 square feet
9. 3 square feet
10. 308 square inches
11. 216 cubic inches
12. 21,952 cubic inches or 13 cubic feet
13. 144 cubic inches
14. 1,163 cubic feet
15. 525 cubic inches
16. 302 cubic inches
17. 17 cubic feet or 28,953 cubic inches
18. 169 cubic inches

Problems

19. 364 square feet
20. 153.3 square feet
21. 92 square feet
22. 371.8 square feet or 379 square feet
23. 241.5 square feet
24. 5 fixtures
25. 302 watts
26. 38 sheets
27. 46 bundles
28. 364 1/4 square inches
29. 981 square feet
30. 402 1/2 square feet
31. 438 7/16 square feet
32. 9,000 square feet
33. 16 square feet
34. 7.07 square feet
35. 615.8 square inches
36. 502 square feet
37. 1,380 square feet
38. 1,321 square inches
39. 12 cubic feet
40. 12.8 cubic yards
41. 5 1/3 cubic feet
42. 6,000 cubic inches or 3.5 cubic feet
43. 13.5 square feet

Chapter 11

Pythagorean Theorem

1. 7.2 inches
2. 8.7 feet
3. 4 inches
4. 12.5 inches
5. 10.4 feet

Problems

6. 31.0 feet
7. 40.5 feet
8. 13.4 feet
9. 13.7 feet
10. 21.1 miles
11. 22.7 miles
12. 19.8 feet
13. 5 feet
14. 6 feet
15. 16 feet
16. 6 feet
17. 50 feet
18. 50 feet
19. to the left

Chapter 12

Operations

1. 14
2. 6
3. 2 5/8
4. 9
5. 10.4

Problems

6. 174 amps
7. $35.03
8. 1 carton
9. 345.6 cubic yards
10. 8,775 square feet
11. See figure below
12. 2.7 gallons
13. 511 hours

Chapter 13

Matching

1. f
2. k
3. l
4. a
5. g
6. e
7. d
8. b
9. i
10. j

ESTIMATE

ABC ELECTRIC
250 WASHINGTON AVE.
ANYPLACE, TX 12345

DATE: __________
SALES PERSON: __________
NO. 2404

EQUIPMENT & PARTS		LABOR			
DESCRIPTION/PART NO.	PRICE	DESCRIPTION	HOURS	RATE	AMOUNT
circuit breaker	4.99	wire addition	7.5	24.25	181.88
150' electrical cable @ $0.29	43.50				
box of staples	2.25				
9 sw. & out. boxes @ $1.39	12.51				
1 oct. box	1.99				
2 S.P. switches @ $2.65	5.30				
7 receptacles @ $1.39	9.73				
2 light fixtures @ $40.00	80.00				

EQUIP. & PARTS TOTAL	$160.27
TAX	$ 9.62
LABOR TOTAL	$181.88
TOTAL ESTIMATE	$351.77

11. n
12. h
13. m
14. o
15. c

Multiple Choice

16. C
17. A
18. A
19. B
20. D
21. B
22. C
23. D
24. D
25. D
26. B
27. D
28. D
29. C
30. A
31. B
32. D
33. C
34. C
35. A
36. D
37. B
38. A
39. C
40. A

Chapter 14

Matching

1. f.
2. c
3. e
4. h
5. g
6. a
7. d
8. b
9. j
10. i

Safety Rules

11. instruction
12. double insulated
13. safety glasses
14. loose fitting
15. guards
16. unplug or disconnect
17. quick-disconnect coupling
18. GFCI
19. flammable
20. ear plugs or hearing protection

Multiple Choice

21. A
22. D
23. A
24. B
25. C
26. D
27. C
28. A
29. B
30. D
31. B
32. C
33. A
34. D
35. D

Chapter 15

Nail Identification

1. common nail
2. box nail or sinker
3. finish nail
4. casing nail
5. duplex nail
6. roofing nail
7. masonry nail
8. staple
9. brad
10. cut nail

Matching

11. c
12. h
13. d

14. a
15. i
16. g
17. b
18. e
19. f
20. j

Multiple Choice

21. D
22. A
23. C
24. C
25. D
26. C
27. B
28. A
29. B
30. C
31. D
32. C
33. A
34. D
35. A

Chapter 16

Multiple Choice

1. D
2. D
3. A
4. C
5. B
6. A
7. D
8. A
9. D
10. D
11. D
12. A
13. B
14. B
15. C

Matching

16. c
17. f
18. a
19. j
20. b
21. i
22. d
23. h
24. e
25. g

Chapter 17

Sketching

1.

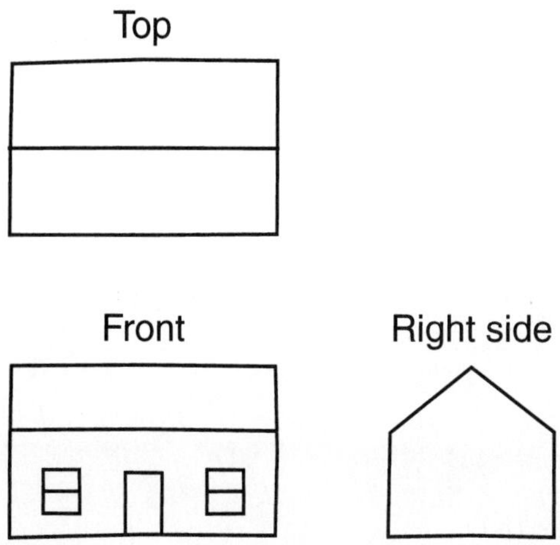

2.

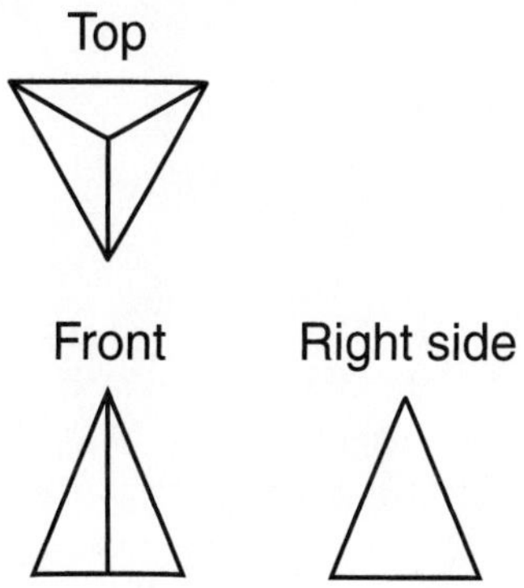

3.

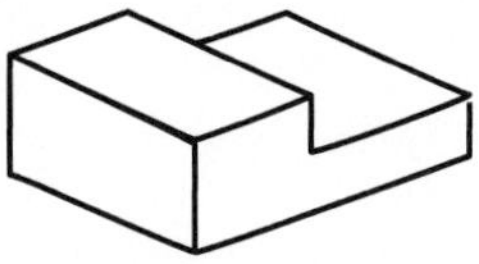

4.

5.

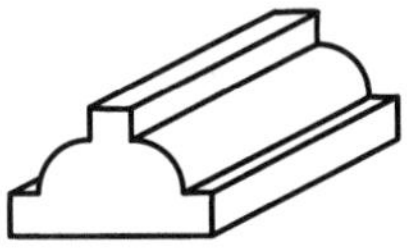

Multiple Choice

6. C
7. D
8. C
9. B
10. A
11. B
12. A
13. C
14. D
15. A

Chapter 18

Scale Use

1. 2′-0″
2. 2′-9″
3. 3′-10 1/2″
4. 7′-0″
5. 10′-9″
6. 0′-7 1/4″
7. 2′-10 1/2″
8. 2′-11 3/4″
9. 1′-3″
10. 5′-10″
11. 5′-3″
12. 6′-8″
13. 7′-11″
14. 1′- 1/4″
15. 1′-2 1/2″
16. 0′-6 1/2″
17. 0′-8 3/8″
18. 1′-1″
19. 2′-4″
20. 2′-10 1/2″

Multiple Choice

21. C
22. B
23. D
24. A
25. B

Short Answer

26. 1/4″=1′-0″
27. 1/4″=1′-0″
28. 3/4″=1′-0″
29. 1″=1′-0″

Chapter 19

Multiple Choice

1. B
2. A
3. D
4. D
5. A
6. C
7. B
8. C
9. D
10. A

Identification

11. Object line. It shows the outline of the building.
12. Dimension line. It shows the length of the building.
13. Dimension line. It shows the depth of the shop.
14. Phantom line. Shows where overhead doors are in open position.
15. Cutting plane line. Shows where a building section is drawn. It is viewed as if facing to the left.
16. Hidden line. Shows concrete footing hidden by earth
17. Break line. Indicates that building continues beyond what is shown.
18. Centerline. Indicates that dimensions are to the centers of the ladder rungs.
19. Extension line. Indicates what dimension line is for.
20. Leader. References a dimension or note to a particular object.

Chapter 20

Matching

1. a
2. c
3. f
4. j
5. d
6. i
7. e
8. g
9. b
10. h
11. m
12. n
13. t
14. k
15. s
16. q
17. l
18. o
19. p
20. r

Application

21. A. north arrow
22. B. brick
23. C. Section B
24. D. double-hung window
25. E. wood

Abbreviations

26. building
27. concrete
28. double
29. ditto (repeat or copy)
30. elevation
31. equal
32. exterior
33. finish or finished
34. floor
35. footing
36. horizontal
37. interior
38. on centers
39. plywood
40. reinforced or reinforcement
41. right of way
42. shower
43. standard
44. tongue and groove
45. typical

Chapter 21

Short Answer

1. 134.92 feet
2. 18 feet x 25 feet
3. 21 feet
4. 101.2 feet
5. 36′-0″ x 60′-0″
6. 9′-0″
7. 2′-6″
8. 3′-6″
9. 8
10. bifold
11. sliding
12. shop
13. 30′-0′
14. East
15. 3

Multiple Choice

16. A
17. B
18. C
19. C
20. A
21. D
22. B
23. D
24. A
25. B

Chapter 22

Short Answer

1. 8′-0″
2. Because of the floor framing
3. mission ‘S’ roof title
4. 1/4 inch in 12 inches
5. 4
6. 32 inches on centers
7. footing
8. 2

9. 1″ x 6″ horizontal trim
10. concrete

Chapter 23

Short Answer

1. This is probably drawing number N on sheet number 4.
2. brick
3. 1/2 inch gypsum board
4. 16 inches on centers
5. 1 inch minimum
6. 1′-0″
7. foundation plan
8. 6 inches
9. insulation
10. 18 inches
11. 3′-0″
12. 1 inch by 1 inch square steel tube
13. 1/4 inch lag screws through 3/4 inch sleeves
14. 1′-1″ on centers
15. 1 1/4 inch diameter steel pipe